ELECTRIFIED SHEEP

Also by Alex Boese

ELEPHANTS ON ACID

HIPPO EATS DWARF

Alex Boese

Electrified Sheep

Glass-Eating Scientists, Nuking the Moon, and More Bizarre Experiments

Thomas Dunne Books
St. Martin's Press
New York

THOMAS DUNNE BOOKS.
An imprint of St. Martin's Press.

ELECTRIFIED SHEEP. Copyright © 2011 by Alex Boese. All rights reserved. Printed in
the United States of America. For information, address St. Martin's Press, 175
Fifth Avenue, New York, N.Y. 10010.

www.thomasdunnebooks.com
www.stmartins.com

ISBN 978-1-250-00753-7

First published in Great Britain by Boxtree, an imprint of Pan Macmillan,
a division of Macmillan Publishers Limited

First U.S. Edition: June 2012

10   9   8   7   6   5   4   3   2   1

To Beverley

CONTENTS

INTRODUCTION

In May 1932, the Polish researcher Tadeusz Hanas wrapped himself in the skin of an ape and crawled into the primate quarters of the Warsaw Zoo. For the next eight weeks, with the permission of the zoo authorities, he lived there, in the company of a group of apes. He ate their food and attempted to learn their language. When he emerged in June, undoubtedly smelling a little ripe, he faced a crowd of curious reporters. 'What was it like?' they asked him. 'What did you learn?' He straightened himself up, tried to look as serious as possible, and replied that he had discovered the habits of apes to be extremely similar to those of men.

I feel a sense of kinship with Hanas. Like him, I've been living with some pretty strange companions. It's been a little over a year since I closed the door of my office and took up residence with them, and now I emerge into the glare of the sunlight, blinking, looking around a little dazed, to present you with this book, which is the result of my year among them.

Unlike Hanas, my companions weren't apes. They were mad scientists. Johann Wilhelm Ritter was one of them. He was a German physicist who systematically electrocuted every part of his body to investigate what it would feel like. Then there was Frederick Hoelzel, who carefully measured the rate at which knots, glass balls, and steel bolts passed through his digestive tract. Mary Henle, the psychologist, liked to hide beneath beds in a women's dormitory and eavesdrop on the conversations of unsuspecting students. I also spent some time with men in Chicago who studiously bent over their desks, drafting plans to nuke the moon.

This book tells about the exploits and bizarre experiments of all of these scientists, as well as quite a few more, and the lengths such people will go to in the advancement of science.

As you've probably figured out, when I say that I lived with these scientists, I don't mean that in a literal sense. They were my intellectual companions. I tracked their exploits through journals, newspapers, and books. Though at times, as I was sitting in the library or in front of my computer screen, it felt like they were there beside me.

Elephants on Acid

Four years ago, I wrote a book called *Elephants on Acid*, also about weird science. The book you're now reading continues the chronicle of eccentric experimentation I started there. I couldn't resist the tug of the mad scientists, pulling me back to spend more time with them.

This time around, I lingered for a longer time with the subjects and characters. Once they got under my skin, I wanted to tell more of their story, but otherwise the structures of the two books are similar. For instance, I again preface each section with a dramatization inspired by a key scene in the story. In these prefaces, I tried to stick as close as possible to real life, though usually I had to make an educated guess at how events may have transpired, and I took some artistic liberties, such as adding dialogue, to heighten the drama. Once you arrive at the main body of each story, however, everything you'll read (to the best of my knowledge) is absolutely true . . . no matter how odd any of it might seem. You'll find a collection of references for further reading at the back of the book, should you wish to delve deeper into any topic.

Because I designed *Electrified Sheep* to complement *Elephants on Acid*, I avoided discussing experiments that I already described in the previous work. So if you're wondering why such-and-such famous weird experiment is missing from these pages, it may well be because it's in *Elephants on Acid*. There are, however, several occasions in which I touch on topics explored in the earlier book. For instance, I look again at vomit-drinking doctors and the home-rearing of apes. In each case I felt it was worthwhile to revisit these topics since there was so much strangeness left to tell about them, and even more bizarre characters to introduce. But apart from these overlaps, which occupy no more than a few pages, *Electrified Sheep* represents a journey into entirely new realms of scientific weirdness.

The Allure of Mad Scientists

So what is it about mad scientists that compelled me to write two books about them? It's definitely not because they're such charming company. Some of them, in fact, are downright creepy.

Part of the appeal of mad scientists is their unabashed freedom to explore new ideas, no matter how wrong such ideas may seem to the rest of us. Most of us have odd notions that pop into our heads from time to time, but we don't act on them. Instead, we go on with our lives without attracting too much attention. We try to stay within the bounds of what's considered normal behaviour. The mad scientist, by contrast, gives free rein to all the strange thoughts swirling around in his grey matter. If he's curious about what it would feel like to cut open his stomach and remove his own appendix, he goes ahead and does it.

Not only do mad scientists naturally and unselfconsciously act on their bizarre impulses, they often get applauded for doing it. If a random person did half the things discussed in this book, they'd either get hauled away to prison or be sent to a mental asylum. For instance, in the chapter on electrical experiments I discuss the British scientist Thomas Thorne Baker, who placed his young daughter in an electrified cage in an attempt to stimulate her growth (although she was of perfectly normal height for her age). If a non-scientist mentioned to his neighbour that he kept his daughter in an electric cage, the police would be knocking at his door in minutes, but because Baker was a scientist, society treated him respectfully, even reverentially. It's only with the benefit of hindsight that we start to wonder, 'What in the world was that guy thinking?'

The exploits of mad scientists also reveal a darker, more emotional side of science, which I find more interesting than scientific stories where brilliant geniuses arrive effortlessly at all the right answers. Here the researchers fumble around blindly as they advance into unfamiliar territory, hoping they're moving in the right direction, but often stumbling down dead-end paths – such dead-end paths, for mad scientists, taking the form of cages in which they sit alone in the middle of the African jungle, or dismal garrets where they lie in the darkness, electrocuting their wasted bodies.

The term 'mad scientist' has a pejorative connotation, but I've actually developed a lot of empathy for some of the characters you'll encounter. It's not my intention to snicker at them. I think science needs its madmen (the less sadistic ones, at least), because it takes a kind of madness to stick to your ideas in the face of difficulty and opposition. Sometimes no one believes you. Sometimes this goes on for years and years. Sometimes you go hungry or, like Sigmund Freud, get addicted to cocaine. Sometimes you perform self-surgery and your guts spill out. But many of the greatest advances in science have been due to stubborn researchers who refused to abandon their theories despite such hardships. Of course, if their theories prove out, they're celebrated and called mavericks, but in actuality aren't mavericks and madmen really two sides of the same coin?

What Kind of Weird?

Four centuries of scientific research has generated a huge amount of weird experimentation. The challenge when writing this book, therefore, wasn't to find enough weird experiments to write about, but rather to choose just a few stories from an embarrassment of riches. So how did I decide what made it into the book?

To be honest, no rigid set of criteria guided my selection process. The absurd, almost by definition, defies easy categorization. I suppose it was the unexpected that I searched for, experiments that joined together unlikely elements, such as psychoneurotic goats and atomic bombs, and therefore seemed to require some kind of explanation. But as I flipped through old journals, I never knew exactly what I was looking for. I waited to be surprised. However, I did know what I wasn't looking for, and that deserves a few words of explanation.

My first rule of exclusion was that anyone trying to be weird wasn't weird enough for this book. Researchers sometimes self-consciously adopt the guise of mad scientists, in a rather slapstick way. They come up with formulas to describe the perfect cheese sandwich, or calculate how long it would take the Large Hadron Collider in Switzerland to defrost a pizza. Science of this kind can be very funny, but it wasn't what I was looking for in this collection. I wanted mad scientists who were totally serious about their offbeat investigations. To my mind, that seriousness of intent makes their behaviour far stranger and more intriguing. In some of the more outlandish experiments, they're not just harmless and wacky. There's a hint of menace lurking beneath their weirdness. Just how far would they be willing to go, we have to wonder, to get the answers they want?

My second rule of exclusion was that I wasn't willing to explore the darkest depths of barbaric acts done in the name of science. Many people, when they think of weird science, imagine scientific horrors – Nazi research, the Tuskegee syphilis experiment, the Imperial Japanese Army's Unit 731. Such things might be strange, and they definitely offer frightening examples of what can happen when all moral limits are abandoned, but to include them would have risked turning the book into a catalogue of atrocities, which I didn't want to do. So I avoided experiments that tilted too far towards the monstrosity end of the ethical scale.

But having said that, I should offer fair warning. Although I don't dwell upon horrors, nor do I offer a pastel version of science. There's a mixture of dark and light in the following pages and there are still experiments that some people might be surprised were permitted in the bona fide pursuance of science. Some of the animal experiments, for example, could appear cruel when viewed with a twenty-first-century sensitivity. But without endorsing any of these more morally opaque experiments, I think it's possible to recognize elements of the surreal and bizarre in even the most grim situations, and that's the spirit in which I offer these stories – as modest reminders that tragedy, comedy, absurdity, and the quest for knowledge are often inextricably intertwined.

Now on to the tales!

Alex Boese

CHAPTER ONE

Electric Bodies

In the ghostly subatomic world, an electron flickers in its orbit around an atomic nucleus. Then, from out of the empty void beyond the atom, comes the distant tug of an attractive force. The electron twitches and leaps, spanning distances billions of times greater than its own size until it comes to rest again around another atom. Out of such microcosmic forces emerges the human-scale phenomenon of electricity. We feel it as an electric shock that jabs our finger, or we see it as lightning in the sky. The mastery of electricity has been arguably the most important achievement of modern science. Our lives today depend in so many ways upon access to electrical power that it would be difficult to imagine survival without it. But the history of electrical research hasn't been limited to the search for technological applications of this force. The urge to understand electricity has been matched by a desire to find increasingly spectacular and unusual ways of displaying its power. In particular, researchers have demonstrated an enduring fascination with exploring the dramatic effect of electricity on living bodies.

Electrifying Birds

Philadelphia, Pennsylvania – 23 December 1750. The turkey eyes Benjamin Franklin suspiciously from across the room. Franklin – middle-aged, balding, and slightly plump – makes a final inspection of his electrical apparatus as a group of men standing behind him watch with interest. The equipment consists of several six-gallon glass jars wrapped in tin. Metal rods protrude from the necks of the jars, and copper wire connects the rods. Franklin straightens up and nods his head appreciatively. As he does so, the turkey, tied to the leg of a table, looks from Franklin to the men and clucks apprehensively.

'The bird isn't happy,' one of the men says.

'But I'll be happy when she's in my stomach,' Franklin replies, and everyone laughs. 'You'll see. Fowls killed by the electrical shock eat uncommonly tender.'

'Then hurry, by all means,' another man says. 'I'm ready to eat!'

At this, the turkey clucks again.

Franklin smiles. 'Soon enough. We're almost prepared. The jars are charged. The only ingredient we lack is the bird.' They all turn to look at the turkey, which stares back at them warily.

Franklin picks up a chain lying on the ground and holds it up for the men to see. 'This chain communicates with the outside of the jars. We need to attach the loose end to the turkey. Philip, if you would, could you fetch the bird here?'

Philip separates himself from the group and walks over to the turkey. He unties it, and, using the rope around its neck, pulls it towards Franklin. The turkey clucks indignantly.

'Good. Now hold it by the wings so I can wrap the chain around its thigh.'

Philip pulls the bird's wings back behind its head, and Franklin kneels down in front of it. The bird glares angrily at him. Just at that moment Franklin's wife, Deborah, enters the room.

'My word, gentlemen. Are you still occupied with that turkey?'

Franklin, distracted, stands up, the chain still in his hand. 'Almost finished, my dear.'

'Well, hurry. The fire is roaring nicely. We shall want to get the bird roasting soon.'

'As soon as we have electrified it, I shall remove its head and bring it in to be plucked and dressed.'

'What times we live in!' one of the men says. 'To celebrate the Christmas season with an electric turkey.'

'And roasted by the electrical jack!' another adds.

The group continues to converse. Keeping one eye on the turkey, Franklin listens to the friendly banter. There's a burst of laughter, and he looks over and chuckles. Absent-mindedly, he reaches his free hand out towards the copper wires connecting the jars. Suddenly there's a flash and a loud bang like the firing of a pistol.

'AAAArrrggh!' Franklin cries out and staggers back several feet before collapsing to his knees. His arms and chest start to tremble violently.

'Ben!' Deborah cries out. The men rush around Franklin to stop him from falling over as convulsions shake him.

'He touched the jars. Took the full discharge!' Philip says.

'Give him air to breathe,' another man orders.

Deborah pushes through the crowd and grasps her husband, holding him as the seizure gradually subsides. 'Ben! Are you all right?'

Franklin looks up at her, stunned, his face ashen.

'Ben, speak to me. Are you hurt?'

His eyes are glassy, unfocused.

'Ben!' Deborah cries out again.

He opens his mouth but no words come out. 'What . . . What happened?' he finally gasps.

At that moment, as if in reply, the turkey emits a loud, self-satisfied cluck.

During the first half of the eighteenth century, an electrical craze swept Europe. Experimenters discovered they could manipulate inanimate pieces of matter, such as glass rods and metal poles, to produce all kinds of spectacular electric effects. Charged objects mysteriously attracted feathers and small pieces of paper. Sparks flew from fingertips. Electrical fire ignited alcoholic spirits and gunpowder. Crowds flocked to see the latest stunts – back then, this was great entertainment – and experimenters competed to dream up ever more dazzling demonstrations. Out of this electrical enthusiasm developed the electrified world we live in today, with our televisions, computers, and brightly lit homes. The story of the birth of the science of electricity has been told many times before, but what people seldom appreciate is the contribution made by the unsung heroes of the eighteenth-century electrical revolution: the birds. These creatures – chaffinches, sparrows, chickens, turkeys, etc. – had the bad luck to be the favourite research animals of early electricians.

The Amazing Electrified Flying Boy

The first electrical experiment truly to capture the imagination of European audiences was Stephen Gray's 'charity boy' demonstration. A young orphan hung suspended from the ceiling, his charged body producing sparks and attracting objects such as brass leaf and pieces of paper. But there, at the beginning, was also a bird, since Gray conducted the same experiment on a 'large white cock'.

Gray worked most of his life as a fabric dyer, but by the 1720s he was a retiree living at London's Charterhouse, a home for down-on-their-luck gentlemen. Being a fabric dyer hardly made him a gentleman, but during his career he had struck up friendships with members of England's Royal Society, with whom he shared an interest in science. These friends used their influence to secure him a place at the Charterhouse, and with little else to occupy his days there, he decided to conduct electrical experiments.

Almost nothing was known about electricity at the time, except that if you rubbed certain substances, such as glass or amber, they acquired the ability to attract light materials – feathers, small pieces of paper, chaff, etc. Substances that acquired an attractive power when rubbed were known as 'electrics'. The term came from the Greek word for amber, *elektron*. Intrigued by these electrics, Gray sat in his room desultorily rubbing a glass tube and picking up feathers with it. But he soon noticed something strange. When he put a cork in the tube, it too acquired the ability to attract feathers. Somehow the attractive power had been transmitted from the glass to the cork, even though cork, on its own, was not an electric.

Realizing he was on to something, Gray explored how far he could transmit this 'electrical virtue'. He inserted a metal rod into the cork, tied packing thread to the rod, and secured a kettle to the end of the thread. Amazingly, the kettle now also attracted feathers when he rubbed the glass tube. He searched for other objects to electrify and found the trick worked on a fire shovel, a silver pint pot, and an iron poker, among other things. Intrigued, he cast his net even wider. The Charterhouse was full of old men. They seemed like problematic subjects for electrification, but then an orphan boy wandered in.

Gray fashioned a harness out of silk and hung the 47-pound boy from the ceiling of his room, parallel to the floor. The boy held his arms out. Gray rubbed the glass tube and touched it to the boy's bare foot. Dust motes and pieces of down floated up towards the boy's hands. To the crowd of old men standing around, the effect must have seemed magical.

Gray shared his discovery with Granville Wheler, a member of the Royal Society, and together they continued experimenting. They strung up Wheler's footboy and learned that if you reached out a finger to touch the electrified boy, you received the prick of a shock. The trick was getting better and better. Curious as to whether the phenomenon only worked with humans, they next tried another species – the white cock. They tied the bird into the silk harness and carefully applied the glass rod. To their delight, the effect was the same as on the boy. The two men circled the rooster, reaching their fingers out towards it as the bird squawked anxiously. They drew sparks from its beak, comb, and claws. The unnamed rooster had become the world's first electrified research animal.

In the interests of science, the two men next killed the bird to determine if its body could still produce sparks. It could. Even plucking it didn't affect this ability. Presumably the researchers concluded that day's work by eating the bird for dinner, though they didn't report that detail to the Royal Society.

For his discoveries, Gray was made a full-fledged member of the Royal Society – a rare honour for a fabric dyer. Word of Gray's flying boy experiment soon spread to the Continent, where aspiring electricians (as electrical researchers were called back then) staged versions of it for delighted audiences in salons and lecture halls. By the end of the 1730s, the stunt had become so popular that it was possible to buy flying-boy electrical kits from instrument makers. These came complete with silk straps and glass rod – like something you'd now buy in an adult catalogue. You had to supply your own boy. Gray's flying rooster had become a mere historical footnote, but electricians had not forgotten about birds. Their interest in them was just warming up.

Bigger Bangs, Leyden Jars, and Eunuchs

Between 1730 and 1745, electrical innovation advanced rapidly. Inventors set to work trying to produce larger amounts of charge in order to achieve even more exciting effects. They replaced the glass rod Gray had used with electrical machines consisting of glass globes or cylinders turned by a crank. Experimenters rubbed their hands on the rotating glass to generate a charge. They discovered that if a metal rod – such as a gun barrel, sword, or empty telescope tube – was hung beside the globe, almost touching it, it collected the electricity, allowing for the accumulation of stronger charges. This allowed for stunts such as the 'Venus electrificata', in which an attractive young woman, electrified via a hidden wire, stood on a non-conducting piece of wax which prevented the charge from escaping to the ground. When a would-be Romeo tried to steal a kiss from her, he felt the tingle of a shock jump from her lips to his.

As the years passed, and the machines grew in power, the subjects of such experiments began to complain that the shocks were actually becoming quite painful. By 1745, electrical machines packed enough of a punch to allow Andrew Gordon, a Scottish Benedictine monk teaching in Saxony, to kill a chaffinch. This bird was the first reported animal killed by human-produced electricity.

The next year, 1746, was an important date in the history of electricity because it marked the invention of the Leyden jar, a device that allowed experimenters to produce, for the first time, shocks of truly formidable power. The instrument took its name from Leyden,

Holland, the place of its discovery, where Pieter van Musschen-broek, a professor, and his friend Andreas Cunaeus, a lawyer, invented it while trying to figure out if it was possible to store electricity in water. Working alone in the lab, Cunaeus ran a wire into a glass jar half-filled with water. Thankfully he didn't have much scientific experience, so instead of keeping the glass jar on an insulated surface, as a competent electrician would have done, he held it in his hand. By doing so, he inadvertently grounded the outside of the glass, turning the jar into a capacitor. When he innocently touched the wire leading into the glass, he created a path between the highly charged interior of the jar and the grounded exterior. The resulting shock knocked him off his feet. He told Musschenbroek what had happened, and the professor tried it himself. He too felt a violent, explosive blow. It was so strong he swore never to repeat the experiment. He urged no one else to try it either.

Of course, that advice was ignored. The Leyden jar astounded scientists. Before that time, electricity had been a mere curiosity, a phenomenon that produced intriguing little sparks and slightly painful shocks. But now, almost overnight, it had become a force strong enough to strike down a grown man. Researchers across Europe scrambled to build their own Leyden jars, and the first thing they did with them was to test the device's killing powers on birds.

The French physicist Jean-Antoine Nollet, who considered himself Europe's leading electrical expert, tested a Leyden jar on a sparrow and a chaffinch simultaneously. He attached the two birds to either end of a brass ruler that had a wooden knob in the middle, allowing him to hold it. He then touched the head of the sparrow to the outside of the jar and the head of the chaffinch to a rod connected with the inside. John Turberville Needham, who witnessed the experiment, wrote to the Royal Society describing what happened next:

> The consequence, upon the first trial, was that they were both
> instantaneously struck lifeless, as it were, and motionless for
> a time only, and they recovered some few minutes after; but,

upon a second trial, the sparrow was struck dead, and upon examination found livid without, as if killed with a flash of lightning, most of the blood vessels within the body being burst by the shock. The chaffinch revived, as before.

Nollet expanded the idea of electrifying a chain of bodies into an even more spectacular demonstration, using humans. As the King of France looked on, Nollet instructed 180 of his majesty's royal guards to hold hands. The man on one end of this human chain touched the rod connected to the interior of a Leyden jar. The guard on the other end waited, and then, when Nollet gave the word, touched the outside of the jar. As soon as he did so, a shock raced through the chain, causing all 180 guards to leap simultaneously into the air. Next Nollet repeated the trick with an entire convent of Carthusian monks. Again, as reported by Needham, 'The whole company, at the same instant of time, gave a sudden spring, and all equally felt the shock.'

A subsequent attempt to replicate Nollet's human-chain experiment yielded an unexpected result. Joseph-Aignan Sigaud de Lafond tried to send a shock through sixty people, but the current consistently stopped at the sixth man. *The man is impotent!* gossips at the king's court tittered. *He blocks the discharge!* Sigaud put it more delicately, suggesting the man couldn't conduct electricity because he didn't possess 'everything that constitutes the distinctive character of a man'. All agreed the phenomenon warranted further testing. So Sigaud gathered three of the king's musicians (all confirmed eunuchs), made them hold hands, and then exposed them to the shock of a Leyden jar. They leapt vigorously into the air! It turned out it hadn't been a lack of virility that blocked the discharge in the original experiment, but rather a puddle the man had been standing in, which directed the current into the ground.

Meanwhile, in Poland, the mayor of Gdansk, Daniel Gralath, built a Leyden jar he used to zap beetles. Then, like Nollet, he killed some sparrows. Curious about the limits of the jar's lethal power, he next tried it on a goose, but at last the jar had met its match. The bird flopped over, as if dead, but soon revived and ran away honking. However, Gralath's experiments weren't fruitless. During the course of his killing trials, he figured out that Leyden jars could be wired together to create shocks of ever greater power, limited only by the number of jars available. He called jars wired together in this fashion a 'battery', because when they discharged their contents the explosion sounded like a battery of cannons going off.

Of course, experimenters weren't shocking birds merely because they thought it was fun. They did so because they had no other way of measuring electrical force. Today we can drive down to a hardware store and buy a voltmeter, but in the 1740s this option wasn't available. The man after whom volts would eventually be named, Alessandro Volta, had only just been born in 1745. So the birds served as a convenient way of approximating force. That is, experimenters could say the force was strong enough to kill a sparrow, but not a goose. It wasn't a very precise form of measurement, but it was descriptive, and it got people's attention.

Not everyone agreed, however, that shocking birds was morally justifiable. Professor John Henry Winkler of Leipzig wrote to the Royal Society in 1746, 'I read in the newspapers from Berlin, that they had tried these electrical flashes upon a bird, and had made it suffer great pain thereby. I did not repeat this experiment; for I think it wrong to give such pain to living creatures.'

Winkler was nevertheless curious about the effects of the Leyden jar, so instead of using a bird he tested it on his wife. He reported that she 'found herself so weak after it, that she could hardly walk'. A week later she seemed to have recovered, so he zapped her again. This time she 'bled at the nose'. The experience was certainly unpleasant for her, but at least no birds were harmed.

Benjamin Franklin vs the Turkey

Across the Atlantic, the electrical experiments delighting Europeans eventually came to the attention of a man who would soon become one of the most famous figures of the eighteenth-century

Enlightenment, Benjamin Franklin. Franklin first saw a demonstration of electrical phenomena in 1743, when he attended a show by an itinerant Scottish lecturer, Dr Archibald Spencer. He was instantly hooked, so he bought Spencer's equipment and began conducting experiments of his own.

Franklin's rise to fame was due, in great part, to his electrical research. Many historians argue that he was, in fact, the greatest electrical scientist of the eighteenth century. It was Franklin who came up with the 'single-fluid' theory of electricity, arguing that electricity was a single force that displayed positive and negative states – terms we still use today. He was also the first to suggest an experiment to prove that lightning was an electrical phenomenon, and, to round off his résumé, he invented the lightning rod. But in the late 1740s, when Franklin first applied himself to electrical research, most European scientists regarded him as little more than a colonial upstart. They believed the most important contribution he could make to science would be to tell them what happened if a large electrical shock was given to that uniquely American bird, the turkey.

Franklin set himself up for the turkey expectations. In 1749, he wrote a long letter to Peter Collinson, a Quaker merchant and member of the Royal Society, excitedly describing his electrical research, most of which involved the systematic investigation of Leyden jars. Franklin ended the letter on a humorous note. Since summer was fast approaching, when electrical experimentation grew difficult because of the humidity, Franklin told Collinson he intended to finish off the season with an electric-themed 'Party of Pleasure' on the banks of the Schuylkill River. The main event of the festivities would be the electrification of a turkey:

> A Turkey is to be killed for our Dinners by the Electrical Shock; and roasted by the electrical Jack, before a Fire kindled by the Electrified Bottle; when the Healths of all the Famous Electricians in England, France and Germany, are to be drank in

Electrified Bumpers, under the Discharge of Guns from the Electrical Battery.

An 'electrical jack' was a kind of primitive electric motor that would be used to rotate the turkey in front of the fire. The 'electrified bottle' was a Leyden jar. 'Electrified bumpers' were electrified glasses, which would give those who attempted to drink from them a shock. And the 'electrical battery' was a group of Leyden jars.

Collinson read Franklin's letter to the Royal Society. They ignored most of it, but the part about electrifying the turkey piqued their curiosity. They asked Collinson to tell Franklin they would be 'glad to be acquainted with the result of that experiment'.

It's not clear whether Franklin had actually been serious about the electric turkey-killing party. His tone suggests his proposal might have been tongue-in-cheek, and there's no other evidence to indicate the unusual banquet occurred. But if Franklin's 'pleasure party' was just a joke, then the Royal Society called his bluff. Franklin now felt obliged to shock a turkey.

Rather than tackling the challenge of electrifying a turkey head-on, Franklin started with hens and worked his way up to the larger bird. First he assembled two large Leyden jars, put a hen in position, and touched its head to the jar. The jars discharged with a bang, and the hen flopped over dead. The experiment had gone off without a hitch, and to his delight Franklin then discovered that the flesh of the bird cooked up 'uncommonly tender'. He speculated this was because the electricity forcibly separated the fibres of the hen's flesh, softening them, though it was actually because the electricity relaxed the bird's muscles and interfered with rigor mortis, which is why poultry farmers today still shock birds before slaughtering them.

Franklin next knocked down a second hen with the Leyden jars, but instead of letting it die he tried to revive it by picking it up and 'repeatedly blowing into its lungs'. After a few minutes, the bird groggily regained consciousness and let out a little squawk.

Delighted, Franklin carefully placed it down on the floor, whereupon it ran straight into a wall. It was alive, but the electricity had blinded it. Nevertheless, this was the first recorded case of the use of artificial respiration to revive an electric shock victim – an accomplishment Franklin seldom gets credit for. People are happy to picture the future founding father of the United States flying a kite in a lightning storm, but giving mouth-to-beak resuscitation to a hen probably doesn't seem as dignified.

Following his success with the hens, Franklin moved on to turkeys. These, however, presented more of a challenge. In fact, in trying to kill a turkey with electricity, Franklin almost killed himself.

It was 23 December 1750, two days before Christmas. A crowd had gathered at Franklin's house to witness the grand turkey electrification. His guests were in good cheer. The wine flowed freely; the conversation was animated. Amid these festivities, Franklin readied two Leyden jars. Finally he called everyone around to see the big event, but by his own admission the merriment of his guests distracted him. He reached out with one hand to touch the top of the jars, to test if they were fully charged, forgetting that in his other hand he held a chain attached to the exterior of the jars. His body completed the circuit. The shock, he wrote two days later to his brother, was like a 'universal Blow throughout my whole Body from head to foot which seemed within as well as without'. His body shook violently. For several minutes he sat dazed, not knowing what had happened. Only slowly did he regain his wits. For several days afterwards his arms and neck remained numb. A large welt formed on his hand where he had touched the jars. If he had taken the shock through his head, he noted, it could very well have killed him.

In the battle of Birds vs Electricians, the birds had scored their first victory. However, Franklin wasn't about to give up. After all, the Royal Society expected results. So, when he had fully recovered, he diligently returned to his turkey experiments, though now with far more caution.

Franklin discovered that two Leyden jars were insufficient to kill a turkey. The birds went into violent convulsions and then fell over, as if dead, but after fifteen minutes they poked their heads up again, looked around, and returned to normal. So Franklin added three more jars to his battery, and in this way succeeded in dispatching a 10-pound turkey. The Royal Society was happy. They congratulated Franklin on being a 'very able and ingenious man'.

Dr Abildgaard's Franken-Chicken

After Franklin, electricians continued to regularly shock birds, but nothing particularly novel was added to such experiments until 1775, when Peter Christian Abildgaard, a Danish physician, reported to the Medical Society of Copenhagen that he had not only killed birds with electricity, but had succeeded in bringing them back to life in the same way.

Abildgaard used hens in his experiment. Employing what was, by now, the established bird-killing technique, he exposed a hen's head to the shock from a battery of Leyden jars. The bird collapsed, seemingly dead. In fact, it really was dead. Abildgaard confirmed this by letting the bird lie there overnight. The next morning it hadn't moved and was stone cold. So Abildgaard tried again with another bird. As before, the bird fell over after receiving the shock, as if lifeless. But this time Abildgaard gave it another shock to the head to see if he could revive it. Nothing happened. He tried again. Still no response. And then yet again. Finally he tried a shock to the chest. Suddenly the bird 'rose up and, set loose on the ground, walked about quietly on its feet'.

Abildgaard was ecstatic. It was the Lazarus of Birds! He was so excited that he immediately killed it and brought it back to life again – not just once, but 'rather often'. After enough of this treatment the hen seemed stunned and could only walk with difficulty, so Abildgaard finally let it be. It didn't eat for a day, but eventually made a full recovery and, to the physician's great delight, laid an egg.

Abildgaard next experimented with a rooster. He shocked it through the head and, like the hen, it fell over, apparently dead. He then revived it with a shock through the chest. The rooster, however, wasn't about to let himself be treated in the same manner as the hen. After returning to life, 'it briskly flew off, threw the electric jar on the ground and broke it'. That was the end of the experiment.

What Abildgaard had discovered was the principle of cardiac resuscitation through defibrillation, though he didn't know this at the time. It wasn't until the twentieth century that doctors realized the full significance of Abildgaard's discovery and electrical defibrillation became a standard part of emergency medicine. To eighteenth-century scientists it seemed instead that electricity contained the power of life itself. Forty-three years after Abildgaard's experiment, Mary Shelley published her famous novel about a mad doctor who used electricity – or, at least, so Shelley strongly implied – to bring a man back to life. If she had been more interested in scientific accuracy, she would have modelled Frankenstein after Abildgaard and made his monster a chicken.

Galvani's Frogs and Tesla's Pigeons

It was the frogs who finally saved the birds from further harm by leaping in to take their place as the preferred research animal of electricians. In 1791 an Italian physician, Luigi Galvani, announced he had discovered a remarkable new phenomenon: 'animal electricity'. The movement of muscles, he declared, was caused by a 'nerveo-electrical fluid' generated within muscles. He demonstrated its existence in frogs, showing how he could make the legs of a dead frog twitch either by exposing them to a spark or by touching them with a pair of metal rods. Galvani's announcement triggered a new electrical craze, of which frogs were the star attraction. The unlucky amphibians were given a place of honour in labs, their bodies poked and probed by researchers eager to summon signs of electrical activity from them. The birds were happy to let them get all the attention.

Although the spotlight shifted away from birds, they didn't disappear entirely from electrical research. Throughout the nineteenth century, occasional reports surfaced of experiments featuring birds. In 1869, for instance, the British physician Benjamin Ward Richardson used the massive induction coil at London's Royal Polytechnic Institute to generate a six-inch spark that he directed at pigeons. They didn't survive.

However, the most prominent reappearance of birds in electrical research occurred during the early twentieth century, and it assumed an unusual, enigmatic form involving the eccentric inventor Nikola Tesla. Tesla was a giant of the modern electrical age. He almost single-handedly designed the technology that made possible the widespread use of the alternating current power that runs through wires in homes today. He then made fundamental contributions to the study of (among other things) high-frequency electromagnetic waves, robotics, neon lighting, the wireless transmission of power, and remote control. It's not an overstatement to say that, without his inventions, the modern world would look very different. But as he aged he developed an obsessive interest in the care and feeding of pigeons. He could frequently be seen around New York City, a thin man in an overcoat and hat, surrounded by huge flocks of birds that he fed from bags of seed. But Tesla didn't merely feed pigeons. He went much further. He believed he had a spiritual connection with the feathered inhabitants of the sky – a connection from which, so he suggested, his scientific creativity flowed.

Tesla spoke of one pigeon in particular – a brilliant white bird with grey tips on her wings – who, for lack of any better term, was his creative soul mate. They spent many years together, but eventually the bird died, and as it did so, according to Tesla, a dazzling white light consumed it, 'a light more intense than I had ever produced by the most powerful lamps in my laboratory'. The bird's death left Tesla feeling lost and aimless. He told a reporter: 'When that pigeon died, something went out of my life. Up to that time I knew with a certainty that I would complete my work, no matter how ambitious my program, but when that something went out of my life I knew my life's work was finished.'

The story of Tesla and the white pigeon is difficult to interpret. The religiously inclined find mystical significance in it. Freudian psychologists read it as Tesla's Oedipal yearning for his mother. Or perhaps it was just the ramblings of a lonely old man. Whatever the case may be, it's curious that a man with an intuitive understanding of electricity as profound as that of any other person throughout history simultaneously developed such a passion and appreciation for birds.

In the present day, birds continue a relationship with electricity that is tense but close, although the electrical utilities are more likely to describe it as an outright war. Every year the utilities spend billions of dollars constructing new transmission lines. The birds respond by raining down excrement on all of them. The white faecal matter oozes its way into delicate insulators causing short circuits that plunge cities into darkness. The utilities send up crews, at enormous expense, to wash the lines clean; the birds drop more poop; and the war goes on. So the next time you're sitting at home reading, or watching television, and the lights flicker and then go out, think of the electrical world the eighteenth-century experimenters bequeathed to us, and then remember the birds.

The Man Who Married His Voltaic Pile

Jena, Germany – February 1802. Outside the clouds shift, revealing the face of the moon. Its light shines brightly through the window of a dark attic apartment, falling on a metallic column that stands on the floor of the room. Caught in the sudden illumination, the column glows like a living creature possessed of an internal source of energy.

The column consists of numerous flat metal discs piled on top of each other. Three tall rods, joined at the top by a wooden cap, press the discs together, as if in a cage, and prevent them from toppling over.

Johann Wilhelm Ritter kneels on the floor in front of the column. He's in his mid-twenties, though years of rough living have aged his delicate features. He wears only a pair of white, ankle-length drawers. The chill of the room has raised goose bumps on the thin flesh of his chest and arms, but he doesn't seem to notice.

His dark eyes flicker with anticipation as he gazes at the column. He runs his hand along the smooth length of it, from top to bottom, caressing it. In response to his touch, the column seems to throb and pulse, glowing, for a moment, even more brightly, though this could just be a trick of the moonlight.

'My dear battery,' he says in a soft voice. 'Are you ready to dance?'

He wets both hands in a bucket of water beside him. Two wires, terminated by metal handles, snake out from the column, one from the top and another from the bottom. Ritter grips the handle of the lower wire with one hand. He reaches for the second wire with his other hand, but before gripping it he hesitates. An expression of doubt, perhaps even of fear, briefly passes across his features but is quickly replaced by a look of iron-willed determination. He grabs the wire.

Immediately he gasps and flinches backwards as if struck by an invisible assailant. The wires don't escape his hands, but he struggles to control them. His arms jerk up and down, fighting with the force that pulses through the wires and into his body. The force twists and bites like a cobra, but finally, slowly, by sheer force of will, he brings it under his control.

Still his hands tremble. The trembling creeps up his arms until it reaches his shoulders.

Both limbs shake now. His lips move, muttering a barely audible prayer, 'Mein Gott, mein Gott, mein Gott.' A line of drool trickles out of his mouth.

For what seems like hours, but is only seconds, he continues to wrestle with the wires. Then, at last, with an explosive motion, he flings them away and collapses backwards onto the floor. He lies there, panting, curled into a foetal position, clutching his arms against his chest. Minutes pass. His breathing gradually calms, and he pushes himself up off the floor.

He gazes at the column, which is still bathed in a pale white glow. 'Quite a lively kick you have, my dear,' he says.

He smiles wryly and hooks his thumbs around the waistband of his cotton drawers, lowering them down, off his slender frame. He pushes *them away and stands fully naked, shivering slightly in the cold air, in front of the moon-illuminated column.*

'*Shall we dance again?*' *he asks.*

Johann Wilhelm Ritter is a name you might encounter in science textbooks. However, you're unlikely to run into his name anywhere else, because, outside of a few obscure academic articles, little has been written about his life. Textbook references to him are often framed discreetly within a sidebar to indicate the information is of historical interest, a small garnish to supplement the weighty prose of the main text. Ritter, you'll be told, is considered by some to be the Father of Electrochemistry, since he suggested, as early as 1798, that chemical reactions can generate electricity. He's also been called the Father of Ultraviolet Light, since he discovered in 1801, through the use of a photosensitive solution of silver chloride, that invisible rays exist beyond the violet end of the spectrum of visible light. A number of other firsts are also attributed to him. He was one of the first to split water into hydrogen and oxygen using electrolysis, and he was the very first to discover the process of electroplating, as well as to build a dry-cell battery and to observe the existence of thermoelectric currents. An impressive list of accomplishments!

Sidebar treatments rarely do justice to any subject, but in Ritter's case the disconnect between such brief biographical treatments and the actual details of his life yawns wider than in most. These accomplishments were only credited to him years after his death, as historians, with the benefit of hindsight, realized the significance of his work. During his lifetime he achieved little recognition beyond a small circle of his ardent supporters. In fact, his contemporaries viewed him as a strange, difficult man – brilliant, but troubled. What he was really notorious for while alive was not any scientific firsts, but rather his bizarre, masochistic methods of self-experimentation with electricity: methods that disturbed his friends and shocked his colleagues.

A Young Dreamer

Ritter was born on 16 December 1776 in the small town of Samitz, Silesia, in what is now modern-day Poland. His father, a Protestant minister, did his best to encourage young Johann to pursue a respectable career, but the boy must have caused him concern. Johann was smart, that was obvious, but he was also a dreamer. He always had his nose in books reading about the strangest things – astronomy, chemistry, and who knows what else. In 1791, when Johann turned fourteen, his father arranged for him to apprentice as a pharmacist in the neighboring town of Liegnitz, but although Johann mastered the necessary skills in no time at all, there were rumblings of complaint from his employer. Couldn't the boy be nicer to the customers? Why was he always so brooding and taciturn? Couldn't he be tidier? The minister feared for his son's future.

If only Ritter senior had known what thoughts were tumbling through his son's head, he would have been far more worried. All kinds of book learning had poured into the boy's mind – science, history, poetry, mysticism – and there they'd swirled together into strange, exotic fantasies. Johann had no interest in preparing lotions and powders to ease the medical complaints of the bourgeois townsfolk of Liegnitz. Instead, he burned with an intense desire to peer deep into the mysteries of Nature. He dreamed of being a scholar, or a poet, steeped in arcane, hidden forms of knowledge. Such ambitions, however, were completely impractical for a minister's son of modest means.

Luigi Galvani's experiments with frogs, which had demonstrated an intriguing link between electricity and the movement of muscles, had particularly inflamed young Johann's imagination. Galvani's work suggested to Ritter that electricity might be the animating fluid of life itself. The same idea simultaneously occurred to many others, for which reason the closing years of the eighteenth century saw researchers throughout Europe busy dissecting frogs and making the limbs of amphibians perform macabre electric dances in their labs.

The form of electricity Galvani had uncovered, a kind that seemed to flow within (and perhaps was created by) bodies, became popularly known as Galvanic electricity or Galvanism, to differentiate it from static electricity. To Ritter, it was a mystery that called out to him. He yearned to know more about it, but as long as he was stuck behind the counter of a pharmacy in Liegnitz, he was powerless to satisfy his hunger for knowledge.

Then fate intervened. In 1795, Ritter's father died, leaving him a small inheritance. Ritter promptly quit his job, packed his bags, waved goodbye to his mother, and took off for the University of Jena in central Germany to fulfill his dreams.

At the time, Jena was an artistic and intellectual Mecca. Poets, scientists, and scholars filled its cafés. It was the perfect place for a young man with Ritter's ambitions. However, when Ritter first arrived he scarcely took advantage of the city's resources. He was so excited by his newfound freedom, and so eager to pursue his galvanic studies, that instead he holed himself up in a rented room with his books and a smattering of scientific equipment (frogs, metal rods, etc.) and began conducting self-guided experiments. There was no separation between his living space and his laboratory. Dishes, dirty clothes, dead frogs, and empty bottles of wine – they all cohabited together. By his own admission, he barely left his room for months at a time since he 'didn't know why he should and who was worth the bother to visit'.

To conduct his experiments, Ritter used the most sensitive electrical detection equipment he could find – his own body. He placed a zinc rod on the tip of his tongue and a silver rod at the back of it. When he did so, he felt an acidic taste, indicating a reaction was occurring. Next he created a circuit that connected his extended tongue to the metal rods and then to a pair of frog's legs. Again he felt an acidic taste, and simultaneously the frog's legs twitched away, proving the presence of a galvanic reaction. He performed similar experiments on his eyeballs (he saw lights dance in his vision), and his nose (he experienced a sharp pain and a prickling sensation).

In 1798, two years after arriving in Jena, Ritter published his results in a book with the wordy title *Proof that a Continuous Galvanism Accompanies the Process of Life in the Animal Kingdom*. At this point, his future looked promising. The book was well received by the scientific community, gaining him a reputation as a skilled experimenter and an expert on galvanism. Professors at the university, such as the famous Alexander von Humboldt, reached out to him to seek his scientific opinions, treating him as a peer, not a student.

Ritter had also finally ventured out of his room and met some of the artists and intellectuals of Jena. At first they didn't know what to make of him. He completely lacked the social skills of the cosmopolitan town's cultured residents. He was really more at home with dead frogs than with people, but there was something about him – his brooding intensity combined with his encyclopedic, self-taught knowledge – that intrigued them. Soon he acquired a reputation as Jena's resident tortured genius and, undeterred by his eccentricities (or perhaps attracted to them), a number of prominent intellectuals befriended him, including the poets Friedrich von Hardenberg (more widely known by his pen name Novalis) and Friedrich Schlegel. This was lucky for Ritter, since he had quickly burnt through his inheritance, leaving him penniless and reliant on handouts from his new friends to survive.

Ritter could never do anything in moderation. His behaviour always went to extremes, and tales of his strange habits became legendary. There were stories about his epic bouts of partying, followed by his equally gruelling stints of complete isolation during which he submerged himself in his work. He was constantly begging for money, and yet whenever he came into funds he spent lavishly on books, scientific equipment, and gifts for his friends. Once he didn't change his shirt for six weeks, until the odour of it became overpowering, and then he wore no shirt at all while it was being laundered. His lack of hygiene was so severe that his teeth started to fall out. And yet, despite this behaviour, he remained incredibly productive, churning out scientific articles that regularly appeared in journals such as Ludwig Gilbert's *Annalen der Physik*.

Even as his lifestyle teetered on the edge of chaos, his scientific reputation was growing steadily.

The Voltaic Pile as Mistress and Bride

In 1800, the Italian physicist Alessandro Volta made an announcement that changed Ritter's life. In fact, it changed the entire direction of electrical research. Volta unveiled a device he called an 'artificial electric organ'. It quickly became more widely known as a voltaic pile, though it did look rather like a tall, phallic organ. It consisted of a vertically stacked column of pairs of silver and zinc discs – or copper and zinc discs – separated by pieces of brine-soaked fabric or paper. The combination of the metals and the brine (the electrolyte) triggered a chemical reaction that produced an electrical current.

If a person placed his hands on the top and bottom poles of this pile of discs, he felt the tingle of an electric current. Stack up more discs, and the current became stronger; the tingle turned into a painful shock. Discs could be stacked ad infinitum, causing the current to become ever more powerful. What Volta had created was the world's first battery, allowing for continuous, steady, and strong discharges of electrical current over long periods of time.

The invention of the voltaic pile opened up numerous new avenues of electrical investigation. Within months of its debut, researchers in England used the device to electrolyse water into hydrogen and oxygen, a feat soon replicated by Ritter. More gruesome experiments were widely conducted on corpses. The bodies of recently executed criminals were transported to surgical theatres, where, under the rapt gaze of audiences, researchers used wires leading from a pile to make the features of corpses twist into horrific grimaces, or caused their limbs to twist and jerk like a marionette.

Ritter immediately fell in love with the voltaic pile. He set to work building his own and busied himself tinkering with it and finding new applications for it. The next two years were the most productive time in his life, as if the pile had energized his intellectual abilities. This was the period during which almost all his 'firsts' occurred, including his discovery of the process of electroplating, his observation of thermoelectric currents, and his construction of a dry-cell battery (a variation of the voltaic pile).

But for Ritter the most exciting aspect of the voltaic pile was that it allowed him physically to experience galvanism. It was like a portal into an invisible world of energy that buzzed and vibrated all around him. He couldn't resist the temptation to plug into that world and discover its secrets, to expose himself to the stinging bite of its current.

Ritter had previously used his body in galvanic experiments, such as when he connected his tongue in an electrical circuit with a dead frog, but the voltaic pile generated far more force. In fact, it was a punishing mistress, though he was willing to endure its lash. The language of romantic play is not simply an affectation. Ritter used it himself. In January 1802, shortly before commencing work on the construction of a massive, 600-disc pile, he wrote to his publisher: 'Tomorrow I marry – i.e., my battery!' His publisher probably didn't realize, at least initially, how literally Ritter meant that phrase.

Ritter began his voltaic self-experiments by stacking between sixty and one hundred discs in the column – an amount that generated a powerful jolt. Then he systematically touched the wires from the pile to each of his sensory organs.

First he clenched both wires in his hands, allowing the current to tingle all the way up to his shoulders. His arm muscles twitched and jerked. He was intrigued by how the positive and negative poles of the pile produced different sensations. For instance, the longer he remained within the closed circuit – sometimes for up to an hour – the more the hand connected to the positive pole grew warm and flexible, whereas the negative side chilled and stiffened, as if exposed to a cold draught.

Next he carefully placed the wires on his tongue. The positive pole produced an acidic flavour – after a few moments his tongue felt as if it were bursting out with welts – whereas the negative pole tasted alkaline and produced an empty feeling, as if an enormous hole had formed in the centre of his tongue. Sticking both wires up his nose caused him to sneeze. When the wires were in his ears, he heard a sharp, crackling buzz on the negative pole and a muffled noise, as if his head was full of sand, on the positive pole. Finally, he touched the wires gingerly to his eyeballs. Strange colours swam in his vision. In one eye, shapes bent and warped. He saw blue flashes. Objects shimmered and bowed outward. In the other eye, everything he gazed at became sharper and smaller, veiled in a red haze.

However, Ritter wasn't done with his testing. There was a sixth sensory organ, that part of the body, as he wrote, 'in which the personal sense of self comes to a peak in its concentration and completeness'. This was his *zeugungsorgan* – his genitals. Ritter was far too thorough a researcher to neglect them.

He waited for darkness to conduct these experiments. Carefully he locked the door. This was not only so that no acquaintance would burst in and catch him in a compromising position, but also, so he said, because he needed to be in a complete state of relaxation to allow him to focus entirely on the interaction between himself and the battery.

His organ began in a state of medium swelling. He wrapped it in a piece of cloth moistened with lukewarm milk – he must have thought milk would be gentler on his skin than brine. Then, delicately, he touched the wire from the positive pole to the cloth, while, with his other hand (moistened for better conduction), he closed the circuit on the other side. A shock jolted him, followed by a pleasant tingling. Not surprisingly, his *zeugungsorgan* responded by swelling. And then it swelled more. The sensation, he admitted, was rather agreeable. Warmth spread out from his groin. Soon he reached a state of maximum tumescence, but dutifully he kept the current flowing. The pleasure built and built, washing over him in waves, until finally – consummation. At this point, he terminated the experiment. He judged it a resounding success.

If Ritter had stopped there, his self-experiments might be remembered as merely a little eccentric, somewhat beyond the pale of normal scientific practice. But given his habit of always going to extremes, he didn't stop. He pushed onward, piling more and more discs onto his voltaic pile – 150, 175, 200. At these strengths, he was able to do serious damage to his body, and he did.

As a result of his brutal self-experimentation, his eyes grew infected. He endured frequent headaches, muscle spasms, numbness, and stomach cramps. His lungs filled with mucus. He temporarily lost much of the sensation in his tongue. Dizzy spells overcame him, causing him to collapse. A feeling of crushing fatigue, sometimes lasting for weeks at a time, made it difficult for him to get out of bed. At one time, the electric current paralysed his arm for a week. But instead of stopping, he merely expressed frustration at the inability of his body to endure more, and despite the difficulty of the experiments, he noted, 'I have not shrunk from thoroughly assuring myself of the invariability of their results through frequent repetition.'

At one point he briefly diversified and dreamed up a punishing experiment that didn't directly involve the voltaic pile. He wanted to test his theory that sunlight was a form of electrical energy, so he decided to compare the experience of staring into the sun with the colours he saw when he placed a wire from the voltaic pile on his eye. With grim determination, he held one of his eyes open, and then he exposed it to the sun for twenty minutes. He stared and stared. A purple dot appeared in his vision. It deepened in colour, and then, after many long minutes, dissolved into a uniform yellow blur. He lost vision in that eye for a month, but before his sight had fully recovered he repeated the experiment with the other eye.

Ritter not only exposed himself to the current of the voltaic pile at greater strengths, but also for longer periods of time. His idea was to use his body as a detector to observe and record the fluctuating cycles of the voltaic pile itself, as if it were a living creature whose strength waxed and waned according to its changing moods. He carefully charted the daily and hourly fluctuations of its power, and as the months passed he was able to create an annual calendar of its capricious temperament, concluding that it grew stronger in the winter and weaker in the summer.

His obsessive relationship with his voltaic pile monopolized increasingly large amounts of his time. Like a lover in a dysfunctional romance, he was always by its side, tending to its every need. However, his lover hurt him, so to ease the pain he self-medicated with alcohol and opium. This in turn fed the self-destructive cycle, allowing him to endure even more time with his metal partner. He noted once in his journal that he had just completed a stretch of five continuous days 'in the battery'.

Ritter had once been the golden child of German science. But when people heard of his self-experiments, they shook their heads disapprovingly. He seemed to have crossed an invisible line past which no one should go, and from which there was no turning back. His behaviour might have been tolerated if it hadn't affected his scientific productivity, but his submissions to journals had grown increasingly incoherent, requiring the heavy intervention of editors to tease out any meaning. 'Never has a physicist experimented so carelessly with his body,' one reviewer later remarked and warned others not to follow his example. Ritter himself noted, with the pride of a masochist, that there was little likelihood of this happening since few people would be willing to replicate the torments he had put himself through.

Recovery and Relapse

In 1803, Ritter had a chance encounter. During one of his rare excursions out of his room, a young woman caught his eye. She was eighteen and beautiful. Her name was Dorothea. Excitedly he wrote to a friend, telling him of the 'girl of delightful quality' he had recently met. The historian Dan Christensen notes that Dorothea was, apparently, a prostitute. Nevertheless, Ritter genuinely loved her. Slowly, under her influence, he began to shake off the voltaic pile's sinister spell. As if rising from the murky depths of a dream, he returned to a semblance of a normal life.

The university administration, however, had grown increasingly unhappy with Ritter. His self-experimentation was disturbing enough, but the faculty had begun to wonder whether he ever planned to graduate. It was now 1804. He had been a student for eight years! It was time, the administrators decided, for him to leave. Recognizing Ritter couldn't afford to pay his graduation fees, they offered to reduce them by half. At first, Ritter resisted. He didn't want to go. He liked his carefree lifestyle. But as the pressure from the university mounted, he reconsidered. Perhaps it *was* time, he thought, to move on and start a new chapter in his life. After all, many of his friends had already left Jena. Perhaps it was time to become a responsible member of the community, now that he had someone he wanted to spend his future with. So he accepted the university's offer. Then, to prove his commitment to a new way of life, he married Dorothea and began looking for gainful employment.

Ritter's self-experiments hadn't completely damaged his reputation. There were still many who thought he had potential, and on the basis of this hope, and his old accomplishments, he landed a position at the Royal Bavarian Academy in Munich. It paid 1,800 gulden per year, which was a fortune to him, and had the added bonus of no teaching requirements. Joyfully the couple packed their bags and headed off to Munich, with a newborn baby to round out the happy scene.

The one portrait that exists of Ritter dates from this period. It's a woodcut depicting a young man dressed in a ceremonial military uniform, made on the occasion of his entry into the Royal Bavarian Academy. In it he looks clean-cut and respectable. His mouth is curved in a slight smile. He probably hadn't looked that presentable in years.

But the old ways weren't so easy to leave behind. In Munich, Ritter found it harder than he had anticipated to change his lifestyle. Little things started to bother him. He disliked the conservative attitudes of the Bavarians. Many of his colleagues didn't share his radical views, nor did they tolerate his eccentricities. Living expenses were more than he had expected. Even with a regular salary, he struggled to make ends meet. Then another child arrived. Ritter grew restless.

A strange new idea began to itch inside his head. What if, he wondered, all the phenomena dismissed by science as supernatural were manifestations of the galvanic force? What if magic was actually a form of electrical interaction between objects? He heard about a young Italian peasant who claimed to be able to detect water and metals beneath the earth with a dowsing rod. Ritter thought it was worth investigating. The twitching of the dowsing rod, he mused, resembled the twitching of a frog's leg in response to electrical stimuli. He petitioned the Bavarian Academy to allow him to travel to Italy to meet this peasant. They were hesitant. They doubted it was real science, but Ritter persisted and finally they relented. In 1806 he departed, full of hope. Once again he was on a voyage of discovery into unexplored new territory.

Ritter returned a year later, bursting with excitement, convinced that the supernatural was a form of electrical activity. 'Here I stand at the entrance to great secrets,' he proclaimed. Eagerly he demonstrated his discoveries to his colleagues, showing them how a hand-held pendulum mysteriously swayed when held over various parts of the human body. His colleagues cast sceptical glances at each other and whispered behind his back: 'What is Ritter up to? If we let him carry on with this, he'll make a laughing stock of the Academy!'

Ludwig Gilbert, the former publisher of many of Ritter's articles, led the attack against him. He published a scathing critique of Ritter's pendulum experiments, dismissing them as pseudo-science, commenting cynically that the only knowledge they could ever produce was knowledge of how the senses can be deceived. Ritter found his colleagues no longer willing to talk to him. He became a scientific outcast.

Chastened, Ritter cast about for a way to redeem himself. In his desperation, he reached out for something that made him feel safe.

Something he knew well. The pendulum had betrayed him, but his old love, the voltaic pile – that had always been true.

His wife must have had reservations when he raised the subject of more voltaic pile experiments. *Not on yourself*, she might have pleaded. But he assured her that he had a different plan. He had always been curious about the presence of galvanism in plants. Now he had a chance to pursue that question. There would be no danger at all.

So the voltaic column moved back in to Ritter's house – like the third member in a bizarre love triangle. Ritter threw himself back into his work. He wrote to a friend that he returned to his experiments *'con amore'*. He spent his days exposing *Mimosa pudica* plants to the stimulation of the voltaic pile, recording how their leaves bent or their stalks twisted in response to the galvanic force. In the long hours he spent with the plants, he began to imagine that they sensed his presence and reacted positively to it – especially, he noted somewhat ominously, if he were sober.

But despite some intriguing results, his colleagues continued to spurn him. His scientific reputation seemed beyond repair. Old aches and pains from his years of self-experimentation also troubled him. To ease them he drank more heavily and increased his opium intake. Debts piled up. He realized he couldn't afford life in Munich with a family.

The breaking point occurred in 1809 when the Napoleonic Wars arrived in Bavaria. The general disruption prompted the Academy to suspend his salary, which hit him brutally hard. With no resources to fall back on, he had no means to support his family. He grew desperate. He didn't know what to do. Finally he sent his wife and children away to live with friends in Nuremberg, while he moved into a small apartment with whatever books and scientific instruments he could carry. He was alone again with the voltaic pile, just like in the old days.

Ritter retreated into the darkness of his room. In December 1809, an old acquaintance, Karl von Raumer, paid him a visit. What he found shocked him.

I came upon Ritter in a vile and dismal room in which everything possible: books, instruments, wine bottles – lay indiscriminately about. He himself was in an indescribably agitated state, full of sullen hostility. One after another he guzzled wine, coffee, beer, and every sort of drink, as though he were trying to extinguish a fire inside of him. . . . I felt the deepest sympathy to see this once so gifted man in such torment, in such bodily and mental suffering.

Ritter was starving. He fell sick with tuberculosis and couldn't drag himself out of bed to beg for food. He wrote to members of the Academy, pleading for their help: 'At noon I'll have nothing to eat, unless some relief shows up.' And then again, 'Please have mercy with me, and please don't be cross, because I call upon you again before receipt of your undoubtedly kind answer.' His letters went unanswered.

On 23 January 1810, a rescue party pounded on his door. 'Ritter! Johann! Open up!' There was no answer. By some means, the door was opened. Holding their sleeves over their noses to ward off the smell, the men entered, picking their way across soiled garments, wine bottles, and scattered pieces of paper. In the midst of this disorder, they found Ritter's body, cold and lifeless, lying sprawled on the bed.

It's not known if Ritter's voltaic column was in the room. Perhaps he had sold it to raise money, but it seems more fitting that he would have kept it to the end, one final link to that magical, galvanic world he longed to explore. If so, we can imagine it there as his rescuers might have seen it – standing guard jealously beside his dead body, gleaming softly and enigmatically in the light that filtered through the shuttered windows.

Electrocuting an Elephant

Coney Island, New York – 10 February 1904. Tony steps out of the workmen's quarters and quietly shuts the door behind him. The night is cold. A wind blows from the east, carrying the scent of the ocean. He lights a cigarette and takes a long drag.

He looks down the avenue towards the amusement park. In the summer, it would be brightly lit, radiating with the glow of thousands of newly installed electric lights. But now, during the off season, it's mostly dark. Above the park, the sky is full of stars. Suddenly Tony tenses. He leans forward, peering into the darkness.

'Madonna mia,' he murmurs. He looks more closely, his eyes opening wide with fright. Then he starts screaming, 'She's back! She's come back!'

The door opens and a dark-haired man pokes his head out. 'Hey, Tony. What's all the screaming about?'

'She's back. Topsy's back!'

The dark-haired man, Jimmy, comes outside, followed by Marco.

'What are you talking about? Who's Topsy?'

'The elephant electrocuted last year! She's back.' Tony points down the avenue, his voice tinged with hysteria. 'Don't you see her? She's right there.'

Jimmy and Marco look down the avenue. Jimmy shrugs his shoulders. 'I don't see nothing.'

Tony's face is pale and his arm trembles. 'We burned her! We did a terrible thing. She's come back for revenge.' He gasps and clutches his mouth, his eyes rolling back in his head as he collapses into a heap on the ground.

'Tony, you OK, man?' Jimmy rushes to Tony's side and looks up at Marco. 'Run and get some help!'

But Marco is staring down the avenue, a look of horror on his face. 'I see her too, Jimmy.' His voice is quiet, almost a whisper. 'It's awful. Her trunk is shooting sparks. Her eyes . . . they're like burning coals.'

'You gone crazy too?'

'Her skin is all on fire. It's terrible. She's real mad. She wants revenge . . . Wait . . . She's fading now. She's fading. I can't see her any more.'

'What's going on with you two?'

Marco puts his finger to his lips. 'Shhh! Listen. Can you hear her?'

'I don't see or hear nothing,' Jimmy bursts out, but then he stops. He did hear something. He shakes his head, as if to clear his ears. It must have been his mind playing tricks on him, he thinks. It was probably just the noise of the surf crashing on the beach several blocks over. But for a second, just a second, he could have sworn he heard the faraway sound of an elephant trumpeting in anger, her cry rising and dying on the night wind.

Topsy the elephant arrived in the United States in 1875 as an involuntary immigrant, imported from India by the circus owner Adam Forepaugh. She was already eight years old when she walked off the ship, but Forepaugh nevertheless promoted her as a baby elephant, parading her down main streets atop the 'gorgeous moving Temple of Juno', a structure 30 feet high. The showman's competitor, P. T. Barnum, pointed out that Topsy wasn't really a baby, but was, rather, a young Asian elephant, a species smaller than its African counterparts. This distinction, however, was lost on audiences for whom all elephants were alike.

Soon Topsy grew too large to pass off as an infant to even the most gullible crowds, and so she no longer enjoyed the status of a headliner. She became just another trick-performing circus elephant. An audience favourite was when Topsy and her companions did the quadrille, a kind of square dance performed by four couples. Their trainer would call out, 'Gentlemen to the right, swing your partners,' and the elephants would lumber through the steps. But even at a young age, Topsy was known for acting up. During the performance, she would often mischievously smack the other elephants on their rumps with her trunk and shower them with sawdust. This penchant for misbehaviour brought unpleasant consequences for her. One time Forepaugh grew so mad at her antics that he struck her on the backside with a stake. The blow broke her tail, gaining her the nickname 'Crooked Tail'.

Life in the circus was hard for the elephants. They were shipped from town to town, kept in small cages, and had to endure screaming crowds and abusive keepers. As the years passed, Topsy soured with age. Finally, in 1900, she snapped and killed two of her trainers. The first one she stomped to death in Waco, Texas. The second one she sat on and crushed in Paris, Texas.

Killing two people earns an elephant a bad reputation, but elephants represent a significant financial investment for a circus, so she was kept on – at least, she was until 1902, when she killed again. Josiah Blount, another trainer, came to wish her good morning. She responded by seizing him with her trunk, throwing him to the ground, kneeling on him, and collapsing his ribcage. At first it wasn't clear why Topsy attacked him, but her other trainers later found a burn mark on the tip of her trunk and speculated that Blount had fed her a cigarette.

Topsy was sold to the owners of the Luna Park amusement park on Coney Island. There she was put to work hauling construction material. Accompanying her was her trainer of many years, Frederick Ault, aka 'Whitey'. He was just about the only person she would listen to, but his presence, in hindsight, wasn't a stabilizing influence, since Whitey was a mean drunk on his own personal path towards self-destruction. Unfortunately, he decided to drag Topsy down with him. In November 1902 the police cited him for animal cruelty for prodding Topsy with a pitchfork, wounding her near the eye. A month later, on 5 December, he led Topsy on a rampage through Coney Island.

Whooping loudly and waving a pitchfork over his head, Whitey rode Topsy down Surf Avenue as a crowd of terrified residents followed at a safe distance behind. Finally she came to a stop in front of the entrance to the exclusive Sea Gate community and refused to go any further. A policeman then led Whitey at gunpoint to the station, but when the inebriated trainer was taken into headquarters, Topsy tried to follow, smashing the front door off its hinges and wedging her entire six-ton body in the frame. Policemen scattered in terror as she began trumpeting loudly and timbers creaked ominously.

Whitey got out of that adventure with a warning in return for leading Topsy back to her stable and promising to behave in the future. But a few weeks later he was back to his old tricks. He released Topsy from her stable, pointed her towards a crew of Italian workmen, and told her to 'Sic 'em.' The workmen fled in terror.

The Luna Park owners decided Whitey had to go, but without Whitey, Topsy was a liability. No zoo would take her, and no other keeper could control her. There seemed to be no other option but to put her down. It was this decision, ironically, that brought Topsy into the history books, because it drew her into a much larger story – a battle being fought between two industrial giants to decide what form of electrical transmission would be used throughout the United States.

The Battle of the Currents

By the end of the nineteenth century, electricity was no longer just a scientific curiosity. It was becoming what we know it as today – an indispensable feature of everyday life. A key factor in this transformation was Thomas Edison's development of a cheap, reliable electric light, which he debuted to an awestruck public in 1879. Of course, an electric light is useless without electricity, and Edison had a plan to provide that too. His company, Edison Electric, opened its first power plant on New York's Pearl Street in 1882, and was soon illuminating much of the city's downtown area. Other cities clamoured for Edison's services. The success of his electrical system seemed assured.

Edison's market dominance, however, didn't go unchallenged for long. He soon faced a powerful rival – George Westinghouse. Westinghouse had made a fortune inventing a new braking system for railroads. The old method had required brakemen to apply brakes individually on each car, jumping between cars to do so. Westinghouse developed a compressed air system that allowed every car to be braked simultaneously. Flush with cash, he decided to enter the electrical business in 1885, having been convinced – largely by Edison's success – that it was too lucrative a market to ignore. Also, he thought he knew a way to beat Edison.

There are two methods of transmitting electricity: direct current (DC), in which power flows in only one direction, or alternating current (AC), in which the current flows back and forth, constantly reversing direction many times per second. Edison's system used DC. This was a proven, reliable technology, but it had one major drawback. It couldn't be transmitted over large distances. The power plant needed to be quite close – within a mile – of the end user. This didn't worry Edison, however, since he imagined each city and town having its own power plant.

Westinghouse decided he could beat Edison with a system designed around AC. The technology was more complicated and poorly understood – Edison believed it was so complicated it could never be made to work – but Westinghouse gambled that if his engineers could get it working, it would prove much cheaper since AC *could* be transmitted economically over large distances. So one power plant could service a huge geographical region. Also, Westinghouse had a secret weapon. He had recently hired a brilliant young inventor, Nikola Tesla, who promised he had solutions to many of the technical challenges posed by an AC system. (The same Tesla, pigeon-lover, that we met earlier.)

So the sides were drawn up in what would prove to be an epic struggle for control of the electrical market. It would be AC vs DC, with Edison's forces lined up behind DC and Westinghouse's industrial might positioned behind AC. Historians describe it as the Battle of the Currents.

The first victory in the battle went to Westinghouse. Tesla delivered on his promise, and by 1888, Westinghouse had built his first power plants and was facing more demand for his business than he could supply. This sent jolts of concern through Edison, as well as everyone whose business relied on DC transmission. Had they attached themselves, they wondered, to a soon-to-be-obsolete technology? But the backers of DC weren't about to go down without a fight. Perhaps DC couldn't best AC on technical merits alone, but there were other, more unorthodox methods of tilting the balance of public opinion back in DC's favour. A letter published in the *New York Post* on 5 June 1888 signalled the start of a new, more bloodthirsty phase of the struggle.

The author of the letter was Harold Pitney Brown, a virtually unknown, thirty-year-old, self-educated electrical engineer. He pulled no punches. 'The alternating current,' he thundered, 'can be described by no adjective less forceful than damnable.' It was true, he conceded, that AC could be operated at lower cost than DC, but by allowing AC wires to run through neighbourhoods, he argued, people were unknowingly putting their lives at risk. An AC wire overhead, he warned, was as dangerous as 'a burning candle in a powder factory'. Someone's death was inevitable. But a DC wire, he insisted, was perfectly safe. Were people really willing to trade lives for a few dollars in savings?

The letter stirred up a hornet's nest of controversy. The backers of AC dismissed his accusations as groundless, but Edison sensed opportunity. Perhaps, he reasoned, Brown was on to something. Perhaps customers could be scared away from AC by convincing them of its dangers. The strategy was worth a try. At the very least, it couldn't hurt to discreetly push a little money in Brown's direction and allow him to be his attack dog. Brown, basking in his new-found notoriety, gladly took the job.

Death-Current Experiments in the Edison Lab

With Edison's encouragement, Brown launched an anti-AC campaign that, to this day, marks a low point in the annals of corporate public relations. Since the public seemed unwilling to accept the danger of alternating current based on his word alone, Brown decided to prove his argument scientifically. He came up with the idea of conducting experiments in which he would test the ability of animals to withstand high voltages of alternating and direct current. The public could then see for itself which current was more lethal.

Brown began his 'death-current experiments' in July 1888 at Edison's corporate laboratory in Orange, New Jersey. He worked late in the evening when the regular employees had gone home. His subjects were stray dogs bought from neighbourhood boys for 25 cents apiece.

Night after night, with grim determination, Brown arrived at the lab, carefully calibrated his equipment, and then proceeded to electrocute dogs. The pitiful howls and cries of the creatures echoed through the building. Scientifically his experiments were little more than a farce, though his results nevertheless appeared in electrical journals. He made no attempt to control for variables such as the weight of the dogs, their physical condition, or the amount of voltage. He simply kept electrocuting dogs until he got the results he wanted, and he ignored any contradictory data.

At times his brutal methods proved too much even for the assistants assigned to help him. One night Brown was giving a 50-pound half-bred shepherd repeated jolts of direct current. The animal had already withstood 1,000 volts, then 1,100, 1,200, 1,300, 1,400, and finally 1,420 volts, but Brown decided to see if it could survive longer exposures, so he gave it 1,200 volts for two-and-a-half seconds. The dog wailed in agony and tried to escape its harness. 'Enough,' an engineer in the room cried out. 'This one's had enough.' He scooped up the dog and took it home. It was one of the few dogs to make it out alive from Brown's laboratory.

After a month, Brown felt he was ready for a public demonstration. He invited electrical experts and members of the press to attend a presentation at the Columbia University School of Mines, at which he promised to illustrate the 'comparative death-dealing qualities of high tension electric currents, continuous and alternating'. On the day of the event, the audience gathered expectantly. Brown walked onto the stage and made some opening remarks. 'Gentlemen,' he announced. 'I have been drawn into this controversy only by my sense of right. I represent no company, and no financial or commercial interest.' This, of course, was a lie. Nevertheless, Brown proceeded to explain that he would prove a living creature could withstand shocks from direct current much better than from alternating current.

Brown's assistants led a 76-pound dog onto the stage. Sensing danger, the dog tried to bite the men as they muzzled it and attached electrodes to its legs. Brown then told the audience he would demonstrate how the dog could easily withstand shocks of direct current. He gave the animal first 300 volts, then 400, 700, and finally 1,000 volts. The *New York Times* reporter, appalled by the spectacle, wrote that the dog 'contorted with pain and the experiment became brutal'. However, it remained alive.

'He will have less trouble,' sneered Brown, 'when we try the alternating current. As these gentlemen say, we shall make him feel better.' Brown gave the dog 330 volts of AC. The weary creature, exhausted by the ordeal, jerked with pain and fell over dead.

There were cries from the audience. Westinghouse's representatives shouted out that the test was unfair. The dog had been half dead already when it was given the alternating current. Brown started to get another dog, but at that moment an agent from the Society for the Prevention of Cruelty to Animals jumped onto the stage, showed his badge, and demanded an end to the experimentation. Reluctantly, Brown complied. As everyone filed out of the lecture room, a spectator was overheard saying that a Spanish bullfight would have seemed like a 'moral and innocent spectacle by comparison'.

Brown's killing spree wasn't over. Four months later, on 5 December 1888, he was back in Edison's lab, having decided to extend his experimentation to even larger animals – two calves and a horse. The calves proved no match for 750 volts of alternating current. The horse, however, presented more of a problem. When Brown first banged the hammer down to close the circuit, nothing happened. The horse looked around, bored. So Brown banged the hammer down again. Still nothing. In frustration, Brown smashed the hammer down repeatedly until smoke rose from the sponge-covered electrodes, but the horse didn't flinch. Carefully Brown checked the connections and tried again. This time it worked. The horse took 700 volts of AC for 25 seconds and collapsed lifeless on the ground.

For Westinghouse, the horse and calf electrocutions, and the publicity they attracted, were the final straw. Enraged, he sent an open letter to the newspapers denouncing Brown's experiments. 'We have no hesitation in charging that the object of these experiments is not in the interest of science or safety,' he declared, 'but to endeavor to create in the minds of the public a prejudice against the use of the alternating currents.'

Brown responded with a brazen proposal. If Westinghouse felt so confident in the safety of alternating current, Brown asked, would he be willing to back up his words with his own life? Would he participate in an electric duel? Brown detailed the conditions of the contest he proposed:

> I challenge Mr. Westinghouse to meet me in the presence of competent electrical experts, and take through his body the alternating current, while I take through mine a continuous current. The alternating current must not have less than 500 alternations per second (as recommended by the Medico-Legal Society). We will begin with 100 volts, and will gradually increase the pressure 50 volts at a time, I leading with each increase, each contact to be made for five seconds, until either one or the other has cried enough, and publicly admits his error.

Westinghouse ignored the proposal, which was just as well for Brown. Westinghouse was a big, husky man – by far the larger of the two – so he probably could have withstood more current.

However, despite Westinghouse's protests, Brown's experiments served their purpose: soon after they helped him score a publicity coup when the State of New York voted to adopt electrocution as the method by which all condemned prisoners would be executed. Influenced by Edison's lobbying, the politicians selected AC as the official 'death current'. In August 1890, William Kemmler, a grocer convicted of murdering his common-law wife with a hatchet, became the first man executed with electricity. Brown and Edison made sure a Westinghouse AC generator was used to kill him.

Electrocuting Elephants

The combination of Brown's experiments and the introduction of the electric death penalty stirred the public's curiosity. Imaginations started to run wild. How much current, people wondered, could other large animals withstand? What would it look like to see them electrocuted? The showman P. T. Barnum, ever mindful of the public's desires, stepped in to satisfy its curiosity. In February 1889, he arranged for electrical engineers to visit the winter quarters of his circus in Bridgeport, Connecticut. There he gave them free rein to electrocute his collection of exotic animals. The engineers made themselves busy. They zapped a seal, hyena, leopard, ibex, wolf, hippopotamus, and several elephants. They didn't use enough current to risk harming the animals; Barnum wasn't about to damage his property. It was just enough to get a reaction. The elephants, in fact, seemed to enjoy the sensation. The *Baltimore Sun* reported, 'They rubbed their legs together, caressed keepers and visitors and squealed with delight.'

This merely whetted the public's appetite. An elephant tickled by current wasn't enough. People wanted more. They wanted to witness the full power of man-made electricity unleashed in the most sensational way possible, in a grim contest of strength with the mightiest creature in Nature. They wanted to see a fully electrocuted elephant.

The dream of electrocuting an elephant was an old one. In 1804, Grimod de la Reyniere, in his *Almanach des gourmands*, wrote of a Monsieur Beyer who owned a giant electrical machine powerful enough to kill an elephant. Grimod was certainly exaggerating, but the fascination with a showdown between man's technology and brute creation was real enough. At the close of the nineteenth century, as workers stretched electrical wires across the countryside to provide the power that would fuel a new industrial revolution, it seemed that man's technology was finally powerful enough to win such a contest. The act would be a symbol of mankind's mastery over the natural world. The editors of *Forest and Stream* magazine, writing in 1896, disparagingly remarked that the American public reminded them of the Roman crowds of old, yearning for 'the spectacle of brute suffering at the hands of man . . . the experiment of killing a vast animal by an untried device.'

Owners of rogue, man-killing elephants were quite willing to provide the public with the spectacle it craved, but for one reason or another, events kept conspiring to deprive everyone of the thrill. An elephant named Chief, advertised as the biggest in America, was the first to be slated for electrocution after he killed three men in 1888. His owners, the circus men John and Gilbert Robinson, announced that electrocuting him would be interesting 'for the novelty of the thing and the scientific possibilities'. But first they granted Chief a temporary reprieve, and then they changed their minds about the electrocution. They eventually used a powerful gun to put Chief down.

Gypsy, in 1896, had a much closer brush with electric execution after she killed two of her keepers. Chicago's Harris Circus applied for a permit to electrocute her, and was ready to sell tickets to the event, but the Humane Society objected to her death being made into a public show. This argument convinced the police commissioner, who denied the permit, citing 'the effect which a public electrocution might have on public morals'. Having failed to electrocute her, the Harris Circus sarcastically offered to ship her to Cuba instead, where, so they suggested, the island's rebel fighters could put her man-killing talents to good use by allowing her to 'trample down the ranks of the Spaniards'.

In November 1901, Jumbo II, who attempted to kill two of his keepers as well as an eleven-year-old girl, got so far as to be wired up with electrodes at Buffalo's Pan-American Midway Stadium. Over 1,000 people paid admission to see him die, but after a last-minute objection by the SPCA – again on the grounds that such a public spectacle was unseemly – the stadium returned everyone's money and sent them home. However, once the crowd was gone, the execution proceeded. An engineer pulled the switch and . . . nothing happened. Jumbo II happily played with a plank. The engineer concluded something was wrong with his equipment, but he was unable to figure out what the problem might be. Eventually Jumbo II's trainers untied him and returned him to his quarters. The *New York Sun* joked, 'He can stand an electric shock of this kind every day.'

It seemed as if people would never have the gruesome satisfaction of seeing an elephant electrocuted. But then, in December 1902, Whitey led Topsy on her rampage through Coney Island, and this time all the planets were in the proper alignment for the public to get what it wanted.

The Death of Topsy

The owners of Luna Park originally planned to hang Topsy, which was a common way to kill circus elephants since it allowed gravity to do the work. They even built a scaffold over the small lake in the centre of the park for this purpose. But the SPCA, as was its habit, objected. For some reason, which went unexplained, the society now felt that electrocution would be more humane than hanging, and it voiced no concerns about the event being witnessed by the public. The green light was given for the electrocution to proceed.

When Edison learned what was to happen, he immediately volunteered to help. By this time the Battle of the Currents was over, and he'd lost. Despite all the scare tactics and electrified dogs, the public had voted with their money, choosing AC over DC. In 1895, Westinghouse had completed construction of a massive power plant at Niagara Falls, which was soon supplying power to western New York. The obvious success of the Niagara Falls plant established, beyond all doubt, the viability and practicality of AC power transmission. So Edison truly had nothing to gain by assisting in the execution of Topsy, but he was still nursing old wounds. In his heart of hearts he honestly believed AC was a death current, unsuitable for anything but killing, and he couldn't resist the opportunity to stick it to Westinghouse one more time.

Edison offered Luna Park the services of three of his top engineers. He also sent along a cameraman to record the entire spectacle. The motion picture camera was Edison's great new invention, the perfection and promotion of which had occupied much of his time during the past decade. He guessed (accurately) that people would flock to view the electrocution of an elephant caught on film.

The only thing Edison didn't send was Harold Brown. By 1903 the two men had drifted apart. Brown was desperately trying to maintain the illusion of a continuing partnership by peddling an invention he called the 'Edison-Brown Plastic Rail Bond'. Edison, however, had sued Brown to stop him from using his name.

The execution was scheduled for 1.30 p.m. on 4 January 1903. A crowd of over 1,500 people gathered in the cold weather to witness the scene. The scaffold over the lake had been converted into an electrocution platform. Several long copper wires snaked out to it.

With all the pomp and circumstance that could be mustered, Topsy's handlers led her out to the platform, but when she arrived at the small bridge that led to the scaffold, she stopped. No amount of force or cajoling could persuade her to go further. A murmur of astonishment rippled through the crowd. And then, as the minutes dragged by, something unexpected happened. The mood of the crowd shifted. They had come to see a death, but Topsy's show of resistance stirred their sympathies. Her behaviour revealed to them an intelligent creature with emotions, deserving of pity. 'Give 'em hell, Topsy!' someone shouted, and a ragged cheer went up.

Worried that the crowd was growing unruly, the Luna Park management sent a message to Whitey, pleading for his help. He replied haughtily that he wouldn't betray his old chum for any amount of money – a noble sentiment, though he hadn't seemed as concerned about her welfare when he led her on the rampage through Coney

Island. Still, the owners were determined to go through with the execution, so they ordered it to be transferred to the open yard. The engineers started to relocate the equipment. They moved quickly, since the crowd was becoming increasingly restless. An hour later everything was ready.

Again Topsy's handlers led her into place. This time she made no protest. At 2.38 p.m. a veterinarian fed her two carrots stuffed with cyanide – insurance to make sure she definitely would die. No one wanted a repeat of Jumbo II's botched electrocution. The engineers attached electrodes that looked like copper sandals to her feet. Then, at 2.45 p.m., they gave the signal to proceed. All of the electrical power for Coney Island, except that needed to operate the trolley cars, had been switched off and diverted into the electrical apparatus. D. P. Sharkey of the Edison Company pulled the switch; 6,600 volts went coursing through Topsy.

The Edison movie camera caught the entire scene in grainy black and white. In film footage that can still be viewed today – do a keyword search on the Internet for the title of the clip, 'Electrocuting an Elephant' – we see Topsy led across the open yard, past stacks of lumber and construction material. The crowd, huddling in their coats, look on from the background. Then there's a jump in the film, and the next frame shows Topsy already standing in place, electrodes attached. She paws the ground with her front foot, as if impatient for the event to proceed. Suddenly she goes rigid. Smoke rises from the ground. She slowly topples over onto her side, and a cloud of smoke completely envelops her. For a few seconds we can hardly see her, but then the wind blows the cloud away, and she lies there, unmoving.

It was all over in ten seconds. Determined to suppress a negative reaction from the crowd, the Luna Park owners immediately began moving people towards the exits and allowed Hubert Vogelsang, a New York merchant, to start carving up Topsy's body. Vogelsang had bought the rights to her remains and hoped to make a profit by selling them off. Her feet became umbrella stands; her organs went to a Princeton professor; and her skin went to the

Museum of Natural History. Edison's film proved to be the most enduring reminder of her existence. In the following years it was shown repeatedly throughout the country, usually bundled together with other short clips such as 'A Locomotive Head-on Collision' and 'An Indian Snake Dance' in order to provide audiences with a full evening of entertainment.

There was nothing new about killing an animal with electricity. Throughout the history of electrical research, almost every variety of creature had fallen beneath the sting of the wires: birds, cats, dogs, cows, horses. Some people had expressed discomfort at these deaths, but few paused to shed any tears, and the animals were quickly forgotten. But Topsy's execution felt different – unusual. Perhaps it was a sense of collective guilt at the sacrifice of an innocent creature of such intelligence to satisfy the whims of industry and popular entertainment. Whatever it was, people found it hard to forget about Topsy after she was gone.

A year after her death, people began to report seeing Topsy's ghost wandering the windswept avenues of Coney Island. Antonio Pucciani, a Luna Park ditch digger, was catching some air late at night outside the workmen's sleeping quarters in February 1904, when he claimed to see her apparition, her eyes blazing with an angry light. Several of his colleagues also witnessed the visitation. The next night a hot-dog vendor saw her, and in the following weeks other staff members reported similar sightings.

It wasn't only humans who were sensitive to Topsy's lingering presence. In 1905, Pete Barlow, trainer of the six new elephants Luna Park had acquired, noticed that his animals refused to walk past a particular spot of ground behind the stables. As they approached it they would trumpet and shake and then come to a halt. Finally he decided to dig there to see what might be troubling them. He discovered the skull of Topsy, buried a few feet beneath the surface. Apparently Vogelsang had left it behind. It was reported that as workmen lifted the skull out of the ground, the elephants trumpeted sadly, and then walked in silence into their quarters.

Two years later, in July 1907, a fire swept through Coney Island.

The fire started in Steeplechase Park, then burned its way up Surf Avenue. Losses were estimated at over $1,000,000. The police determined the cause of the conflagration to be a cigarette discarded in a pile of trash, but other, more sinister theories circulated. Many suspected it was the work of the Black Hand, or Mano Nera, a secretive Italian criminal organization known to threaten business owners with ruin if they refused to meet their financial demands. The more superstitious whispered it was the work of Topsy, come back to exact fiery revenge.

The fascination with Topsy, and sense of regret at her passing, only strengthened as the years went by, continuing right up to the present. The past two decades have seen a steady flow of scholarly articles about her, as well as artistic tributes including short stories by Joanna Scott and Lydia Millet, and songs by indie-rock groups Piñataland and Grand Archives. A high point of this popular interest occurred in 2003, on the 100th anniversary of her death, when artists Gavin Heck and Lee Deigaard joined forces to install a memorial to Topsy at the Coney Island Museum. The memorial allows visitors to stand on copper plates resembling the electrodes used to kill her and watch the film of her death through a 'mutoscope', a coin-operated, hand-cranked viewing device.

It's difficult to find anything worthwhile in Topsy's death. It was a senseless act, designed only to showcase mankind's newfound industrial might by using the power of science to strike down one of Nature's mightiest creatures. If electrical engineers of the time could have figured out how to electrocute a blue whale, they probably would have. However, Topsy's sacrifice did have one positive effect. It seemed to sate the public's urge to view the spectacle of a vast animal killed by electricity. At least no more elephants were ever executed in that manner in the United States – nor, quite possibly, in the rest of the world. Other elephants have died of electrocution, but always accidentally, usually by wandering into power lines. It's a small victory, but certainly the best tribute of all to Topsy.

From Electro-Botany to Electric Schoolchildren

London, England – August, 1912. The first thing the reporter notices as he enters the laboratory is a dark-haired young girl sitting on a stool inside a tall, square cage. She wears a white dress, has a gap-toothed grin, and appears to be no more than five years old. The next thing he notices is the electrical apparatus positioned next to the cage: coils, levers, wires, switches, and other gear whose purpose he couldn't begin to guess.

A man wearing a bow tie and an ill-fitting grey suit approaches. He's no more than thirty years old, slightly built, and has Mediterranean features – dark hair, olive skin, and brown eyes. He extends his hand. 'Greetings. You must be John Sloane, the London Mirror *man. Thomas Thorne Baker at your service.'*

'Pleased to meet you,' Sloane replies as he shakes Baker's hand.

'And in the cage is my daughter, Yvonne. Say hello, Yvonne.'

'Hello, Mr Sloane,' Yvonne says, with a faint hint of a lisp.

'Hello, Yvonne!' Sloane answers, waving to her awkwardly, unsure of the proper way to greet a girl in a cage. She waves back enthusiastically.

Baker immediately takes charge of the interview. 'Let me explain the principle of my experimentation, and then I shall give you a demonstration. My previous work, as you may be aware, has demonstrated the stimulating effect of high-frequency electromagnetic currents upon the growth of chickens.'

'Chickens?' Sloane asks, as he reaches into his jacket pocket for a notebook.

'Yes, chickens,' Baker confirms. 'I've found that chickens exposed to high-frequency currents not only grow faster and larger than their non-electrified counterparts, but are calmer and easier to handle. I reasoned that if it works for chickens, well, why shouldn't it also work for children!'

Sloane nods as he writes in his notebook: 'Works for chickens . . . children also?'

Baker gestures at the apparatus beside the cage. 'To generate the high-frequency currents, I use a coil of the kind designed by the American inventor Nikola Tesla. When I activate the current, the atmosphere inside the cage is saturated with incalculable millions of infinitesimal electric waves moving hundreds of thousands of times a second.'

'And this isn't harmful in any way?'

'Quite the opposite. The high-frequency currents stimulate the circulation by lowering the viscosity of the blood, thereby increasing vitality. I would not use Yvonne as my subject if I were not certain of this.'

Yvonne grins at the mention of her name.

Baker continues, 'Once I have collected enough data with Yvonne, I will be able to offer treatment to babies who are thin, anaemic, or underweight. Many medical men have already expressed an interest in my work.'

Sloane arches an eyebrow. 'I wonder what an old-fashioned mother would say if she could see your method of strengthening babies?'

Baker laughs – a short, bark-like sound. 'What would she say, indeed! It is progress, Mr Sloane. Progress! But please, allow me to give you a demonstration.'

Baker turns towards the electrical apparatus and makes a quick visual inspection of it. 'Are you ready, Yvonne?'

Yvonne nods her head. 'Yes, Daddy.'

'Mr Sloane, I am now turning on the current.' A loud electric hum fills the room. Blue sparks flash from a tightly wound coil of copper wire.

Baker picks up a sealed glass tube and holds it out towards Sloane. 'This is a helium vacuum tube,' he explains, raising his voice to speak over the noise of the equipment. 'Observe what happens when it is placed in the electrified atmosphere.' He moves the tube near the bars of the cage, and as he does so it emits a soft yellow glow. He moves it away and the glow fades. Baker places the tube back on his lab bench.

'What would happen if someone touched the bars?' Sloane asks.

'You are welcome to try.'

Sloane extends his hand, but as his fingers near the metal a spark leaps from the bars to his skin. 'Owww!' Sloane cries, pulling his hand back and shaking it.

Baker laughs again. 'It is not powerful enough to do any damage, I assure you.'

Sloane, holding his injured finger, turns to Yvonne. 'Tell me, Yvonne. How does the electricity make you feel?'

Yvonne considers the question for a moment, then gazes up at the ceiling and replies, with a dreamy expression, 'I feel lovely all the time. It makes me feel very happy.'

'That's wonderful. But are you not concerned at all?'

Yvonne shakes her head. 'No. It is so comfortable. I should like to go to sleep here.' At that moment the electric coil emits a loud, crackling pop. Yvonne adds, as if for emphasis, 'I love having 'lectric currents!'

A flicker of concern passes over Sloane's features. He smiles, a little warily. 'So I can see,' he says. 'So I can see.'

The pursuit of electric growth began during a cold Edinburgh winter at a boarding school for young women. It was December 1746. Stephen Demainbray, master of the school, took a myrtle bush from the greenhouse and placed it in the front room, near the door. He then electrified the plant every day for seventeen days. Unfortunately, we don't know for how long he electrified it each day (one hour? Five hours?), nor do we know the exact method he used – whether he touched wires to the leaves or stuck them in the soil. However, we can guess at the kind of machine he used because the electricity-generating machines of that era typically consisted of glass globes that a researcher spun by means of a crank. He rubbed either his hand or a leather pad against the moving glass to produce a static charge. To produce a continuous charge, one had to keep cranking the machine, and this laborious task often fell to an assistant. In Demainbray's case, the assistant was probably one of his young female students.

Although Demainbray was vague about the details of his experiment, he was clear about the results. In a letter sent to *The Gentleman's Magazine,* he excitedly reported that the myrtle had produced several new shoots, the largest one measuring a full three inches. This was in spite of the cold weather and the gusty draughts in the room as his students opened and closed the door. Myrtles left in the greenhouse, on the other hand, didn't grow at all.

Demainbray concluded that electricity must have a stimulating effect on plant growth. He suggested his discovery could be of great benefit to society 'if the hint be rightly taken'.

Electrical researchers needed little urging to take the hint. In the following decades, they devised all manner of ways to electrify seeds, shrubs, and shoots in search of the elusive combination that would supercharge the growth of plants. They stuck lightning rods in fields, buried electrodes in the ground, and stretched webs of iron wires over crops in order to collect 'atmospheric electricity'.

There were tantalizing reports of success. In 1783, the clergyman Abbé Bertholon claimed his lettuces had grown to a remarkable size when he watered them with an electrified watering can. Even more exciting was the widely circulated, though dubious-sounding, tale of a dinner party hosted by Henry Paget, 1st Marquess of Anglesey. It was said that before Paget served his guests dinner he had them place a few seeds of cress in a pot that contained a mixture of sand, oxide of manganese, and salt. Paget then electrified the pot. By the end of the main course, the cress had supposedly grown to full size and was served to the guests in their salad.

No one knew why electricity should have a positive effect on plant growth, but then the nature of electricity itself was a complete mystery, so confusion was to be expected. Some theorized that the current increased the flow of sap. Others suggested it caused vibrations, thereby loosening the earth. But the most prevalent hunch was that electricity might be the animating force of life itself. Therefore it made sense that, if applied in judicious quantities, it would increase vitality.

Of course, not all researchers agreed that electricity did, in fact, invigorate plants. Many would-be electro-horticulturists were rewarded with nothing but withered flora for their efforts. Others could discern no effect at all. But the true believers didn't succumb to doubt. They were sure that electricity must have a stimulating effect on vegetation, even if some researchers seemed unable to replicate their results.

Dr Poggioli's Electrical Gymnastics

It was in 1868 that the leap from plant growth to human growth was first made. Dr Poggioli, the official physician of the Italian Music Academy in Paris, presented a paper to the French Academy of Medicine titled 'The Physical and Intellectual Development of Youth by Electricity'. Poggioli observed that if electricity quickened vegetable growth, then it might also boost the physical and intellectual powers of children. In support of this theory, the doctor described the case of a boy who had recently come under his care. The child, said Poggioli, had been 'a phenomenon of deformity and stupidity'. But after only one month of electrical treatment the boy had grown an entire inch and had become the top student in his class. Poggioli, like Demainbray before him, didn't elaborate on exactly what this treatment involved – he later referred to it mysteriously as 'electrical gymnastics' – but it seems to have involved the use of a powerful battery to shock various parts of the body. It's safe to assume this was painful.

Poggioli proposed a test of his theory. Give him the bottom six students in any school, let him loose on them with his electrodes, and they would soon, he promised, be among the top in their class – and taller to boot! It's not recorded whether Poggioli ever got a chance to conduct his experiment on any unlucky children, but the next year he was back in the news with a new proposal. He wanted to straighten all the hunchbacks in France. Once again, it was his electrical gymnastics that would do the trick. He estimated there were over 50,000 hunchbacks in the country, so he had his work cut out for him.

Electromagnetic Chicken Coops

After Poggioli, interest in electrical child growth languished for the next forty years. Then, in 1912, the idea experienced a revival. It was the height of what historians call the Golden Age of Electrification – the period when electricity became the driving force of the industrial economy. Transmission wires were being strung throughout the countryside. City streets were suddenly ablaze with electric lights. Electric motors hummed in factories. The public embraced electricity as the symbol of everything modern, new and vital. Its potential seemed limitless and inevitably beneficial. Physicians, jumping on the bandwagon, promised 'electric' cures for everything from baldness to fatigue to impotence. Given this great swell of enthusiasm for electricity, it seemed only natural to wonder whether its positive effects extended also to small children. At least, it seemed natural to the English researcher Thomas Thorne Baker to wonder this.

Baker was a young British electrical expert eager to make a name for himself. He got his first big career break in 1907 when the *Daily Mirror* hired him to help develop a method of sending photographs over telephone lines. He built what he called the electrolytic telectrograph, which could transmit a grainy but recognizable image in about ten minutes. It was a primitive version of a fax machine. However, his invention wasn't the only one of its kind in existence – researchers had been experimenting with the electrical transmission of images since the 1880s – and the *Daily Mirror* soon decided that the operating costs of his telectrograph were too high for it to have much practical use. But meanwhile, Baker's mind teemed with other ideas. He invented an electrical lock that opened in response to specific musical tones. He imagined selling it to chapels so that they could automatically fling open their doors whenever the notes of a wedding processional were played. Then he turned his attention to high-frequency currents, a phenomenon made famous by the brilliant, eccentric inventor Nikola Tesla.

Tesla had made a fortune during the 1880s by inventing much of the technology underpinning the use of alternating current to transmit power – when he was working for Westinghouse, as we saw earlier. With this money, he opened a laboratory in New York, where, during the 1890s, he developed new electrical technologies including an oscillating transformer, or 'Tesla coil', that allowed him to amplify electrical signals to ever higher frequencies and voltages. These high-frequency currents, he found, displayed unusual and dramatic effects. For instance, his coil could send bolts of high-voltage lightning arcing across a room. It also created powerful electromagnetic fields that caused gas-filled tubes to glow brightly, even though no wires were connected to the tubes.

It was these electromagnetic fields that interested Baker. A person felt nothing when standing inside such a field. In fact, the energy seemed to have no effect on living tissue, but Baker suspected this couldn't be true. All that electromagnetic energy saturating the atmosphere must be doing something. He theorized that being washed by invisible electric waves might actually have a positive influence – just as the sun's rays were health-giving. He decided to explore the phenomenon further in the hope of finding a way to make money from it.

Baker started his high-frequency experimentation with peaches and Camembert cheese, since the food industry seemed like the most obvious customer for any discovery he might make. He exposed the fruit and cheese to electromagnetic fields, and both seemed to ripen faster. Encouraged, he soon moved on to a larger, potentially more profitable organism: chickens. In his backyard he built an electrified coop, large enough to house twelve chickens. The birds perched on insulated wires that, for one hour each day, he charged with 5,000 volts. The English correspondent for *Scientific American* visited Baker to examine the electro-coop, and out of curiosity reached out to touch one of the chickens. A spark leapt from the bird's beak to the reporter's hand. The reporter was taken aback, having never been shocked by a chicken before, but oddly the chicken seemed unfazed. Baker told him the chickens actually quite liked the electricity. Whenever they heard the characteristic 'zzzzzz' of the current being turned on, they cocked their heads to one side, as if listening, and then eagerly hopped on the wires.

Baker reported that his electric chickens made spectacular progress. Soon they weighed 13 per cent more than their counterparts in a non-electrified coop. They did seem a little sluggish – stunned, perhaps – but Baker figured this was a good thing. It meant they ate less food and were easier to handle. So he managed to persuade a commercial chicken farmer, Mr Randolph Meech of Poole, to allow him to conduct a full-scale trial on his farm. He wrapped an entire chicken building, housing some 3,000 chickens, in insulated wire, and was soon boasting that the birds inside it grew 50 per cent larger in half the time.

Despite such remarkable results, chicken farmers, being cautious by nature, didn't rush out to electrify their coops – though there was one report of a Brooklyn dentist, Dr Rudolph C. Linnau, who became so excited by what he heard that he gave up pulling teeth and went into the electro-chicken trade instead. Nevertheless, Baker forged ahead, moving on to a larger and even more ambitious subject: young children. He reasoned that what worked on a chicken should also work on a child. His plan wasn't to fatten the children up for consumption, of course, but rather to find a way to treat underweight children by saturating them with health-giving electric energy. He also imagined there might be significant financial rewards if he found a cure for shortness.

In his London laboratory, Baker constructed an electrified cage, through which hummed and pulsed thousands of volts of high-frequency current. A *London Mirror* reporter who interviewed Baker described wires snaking across the laboratory floor, blue sparks crackling from a giant coil, and electrical apparatus that radiated a 'sense of mystery and unknown power'. Into the cage in the centre of this buzzing electromagnetic environment, Baker placed his five-year-old daughter, Yvonne. A photo accompanying the *London Mirror* article showed her seated in the grim-looking contraption. The cage resembled the kind used in courtrooms to house violent criminals, though Yvonne, wearing a frilly white dress, looked the very picture of innocence. She had a slight grin on her face as if she found the entire experience a fun game.

Unfortunately we don't know what results Baker achieved – whether Yvonne shot up in size, or developed remarkable mental powers – because Baker never published any updates about his experiment. Most likely he abandoned it after he realized he'd been upstaged, because at around this time, the summer of 1912, stories began to appear in the press about a similar but far more elaborate study conducted in a Swedish classroom by Svante Arrhenius, winner of the 1903 Nobel Prize in Chemistry.

Electric Schoolchildren

Breathless press reports detailed an elaborate experiment conducted by Arrhenius in which he concealed wires in the walls and ceiling of a classroom, turning it into a giant solenoid, or electromagnet. Twenty-five students and their teachers sat in the room, unaware of the 'magnetic influence' surrounding them. There was, however, said to be a lingering smell of ozone in the air, which must have alerted them to the fact that something strange was going on – that and the constant electric buzz. After six months, Arrhenius compared the students' progress to a similar group in a non-electrified room. The electrified students reportedly scored higher on all counts. They grew almost 50 per cent more and earned higher marks, on average, on their tests. The teachers also benefited from the treatment, remarking that they felt 'quickened' and that their powers of endurance increased.

The Swedish experiment made headlines around the world, but it particularly caught the attention of Tesla in New York, who complained to the press that Arrhenius's electrical apparatus 'was in all essentials the same as the one I used in this city many years ago when I had an installment of that kind in continuous use'. Except that Tesla hadn't electrified any children. But he spied an opportunity in the Swedish study. Since the 1890s, his financial situation had deteriorated. An ambitious effort to broadcast electricity wirelessly around the world had fallen apart, erasing much of his wealth with it. By 1912 he was desperate for new sources of funding, and electric growth, he thought, might be just the thing to restore his finances.

Tesla scheduled a meeting with William Maxwell, the New York superintendent of schools, and urged him to repeat Arrhenius's experiment with American children. He assured him there would be no risk of danger, and the benefits might prove great. Tesla noted he had once employed a rather dull assistant who had developed 'remarkable acumen' under the influence of the electromagnetic fields permeating his lab. Of course, exposure to high-frequency currents hadn't appeared to hurt Tesla's intelligence either.

With the support of Dr Louis Blan, a Columbia University psychologist, and S. H. Monnell, a Chicago doctor, Maxwell agreed to proceed. But he decided the American study would differ in one significant way from the Swedish one. Instead of regular schoolchildren, mentally handicapped students would be used as subjects. His reasoning was that mentally handicapped students were more in need of help. 'Electricity For Defectives', declared a non-politically correct *New York Times* headline. 'The brains of the children will receive artificial stimulation to such an extent that they will be transformed from dunces into star pupils,' the *Times* reporter gushed.

Tesla, who was to be in charge of the experiment, eagerly held court with the media, painting a picture for them of a time in the near future when every home would have its own in-wall Tesla coil. Living rooms, he envisioned, would become electric cages pumping their occupants full of nourishing electromagnetic energy. The *Times* paraphrased his bullish prediction of how this would transform society: 'Ordinary conversation will then be carried on in scintillating epigrams, and the mental life of the average adult will be so quickened as to equal the brain activity of the most brilliant people living before the time when a generator of high-frequency currents was a household essential.'

Everything was ready to go. Tesla had even priced out the equipment. Then disappointing news arrived from Europe. The details of Arrhenius's experiment, it turned out, had been seriously misreported. The British psychiatrist James Crichton-Browne had written to Arrhenius, seeking more information about the study. Arrhenius had responded, informing him that almost all the facts in the newspapers were wrong. He *had* exposed a group of children to high-frequency electromagnetic currents. That much was true. But they had been newborn infants in an orphan asylum. So the claims of boosted intelligence were fictitious. Initially his results had been promising. He had observed a rapid weight gain among the electrified children, but when he examined the study's methodology more closely, he discovered that an overzealous nurse had placed all the healthiest children in the electrified group, and the weakest ones in the control group. Upon repeating the study with stricter oversight, the apparent benefits of electricity disappeared.

Arrhenius's discouraging results took the wind out of the sails of the electric growth movement. Superintendent Maxwell quietly shelved the plans for electrifying schoolchildren. After all, if it didn't work in Sweden, he wasn't going to risk trying it in New York. Tesla went back to looking for other ways to make money, though he focused increasing amounts of time on his true passion: caring for pigeons. Thomas Thorne Baker returned to his work on the electrical transmission of images – work that eventually helped pave the way for the development of television broadcast technology. Arrhenius himself, frustrated in his efforts to improve infants by electrifying them, turned to alternative methods of boosting the stock of Sweden. He became an ardent supporter of eugenic 'racial hygiene' policies and lobbied for compulsory sterilization laws.

Liquid Sunshine and Pigs in Space

Although electric growth cages were assigned to the dustbin of medical history, the underlying thinking that gave rise to them – the conviction that invisible energy rays must have a wholesome influence – proved far more resilient. There was a seductive logic to the idea that being bathed in energy should have an uplifting effect. As the physicians Hector Colwell and Sidney Russ wearily lamented in 1934, 'For some reason or other there appears to be a widespread tendency in the public mind to regard everything connected with "rays" as on that account conducive to health and vitality.' Offshoots of this enthusiasm for invisible energy kept popping up in new settings.

For instance, during the early decades of the twentieth century, as prospects for the health-giving effects of electrical energy waned, enthusiasts transferred their hopes to a new, even more promising phenomenon: Radium Energy! Pierre and Marie Curie first isolated radium in their lab in 1902. This mysterious metal appeared to produce a limitless amount of energy. The Curies noted with amazement how it always remained hotter than its surroundings. If they tried to cool it down, the metal simply heated up again of its own accord, as if in defiance of the Second Law of Thermo-dynamics. It kept pumping out energy month after month, year after year. And where there is energy, medical entrepreneurs noted, there must be health!

Physicians swung into action, promoting the beneficial effects of 'radiumizing' the body to an eager public. Retailers sold radium-treated water, describing the faintly glowing solution as 'liquid sunshine'. The marketers for Hot Springs, Arkansas, whose natural springs were found in 1914 to contain high natural levels of radium, prominently featured this finding in their advertising literature, noting that 'radioactive substances, unlike any other electro-therapy, are able to carry electrical energy deep into the body', thereby invigorating the 'juices, protoplasm, and nuclei of the cells'. Even Thomas Thorne Baker briefly latched onto the craze. He reported to the Royal Society of Arts in 1913 that he had obtained a 400 per cent increase in the size of radishes by growing them in radium-treated soil.

The radium craze persisted well into the 1930s. Marie Curie her-self insisted on the metal's health benefits, maintaining this belief right up until 1934, when she died of overexposure to radiation. It was only the atomic bomb and fears of nuclear fallout that finally cast a permanent shadow over radium's reputation.

Still, a faith in the positive impact of invisible energy has endured in popular culture. The historian Carolyn Thomas de la Peña has noted how revealing it is that fictional comic-book super-heroes frequently gain their powers by exposure to radiation or electrical energy. For instance, Barry Allen turns into the Flash after a lightning bolt shatters vials of chemicals in his lab. A radioactive spider bites Peter Parker, transforming him into Spiderman, and the Fantastic Four gain their abilities after accidental exposure to cosmic rays in outer space. The logic of radiation-powered superheroes can be traced back to beliefs in the benefits of electromagnetic growth therapies and radiumized bodies.

A curious descendant of the invisible energy enthusiasm can even be found in the present day in a rather unlikely place – the Chinese space programme. Chinese scientists, from the very start of their space programme, have expressed great interest in the effect of cosmic rays on plants, hoping that such rays might produce Super Veggies to feed their growing population. At first they used high-altitude balloons to fly seeds up to the edge of space. Now seeds are taken aboard the Shenzhou spacecraft. The resulting crops, grown back on earth, are occasionally served in Shanghai restaurants. Space spuds, it's reported, taste more 'glutinous' than terrestrial varieties.

On 12 October 2005 the Shenzhou VI spacecraft blasted off carrying a particularly special cargo – 40 grams of pig sperm to be exposed to cosmic rays. Whether or not the experiment generated positive results is unknown, because, after the initial announcement, a shroud of official state secrecy descended upon the mission. But maybe, somewhere on a farm in China, a giant, cosmic-ray-enhanced pig is rolling happily in the mud. When Mr Demainbray electrified his myrtle plant back in 1746, he certainly could never have predicted that it might one day yield such a strange progeny.

Lightning, Churches, and Electrified Sheep

Pittsfield, Massachusetts – 1923. The village presents a bucolic scene as it slumbers in the late-afternoon glow. A general store stands at the crossroads. Clustered nearby are a handful of houses. Trees line the side of the road. Further out stands a single church, its spire reaching into the sky.

Beyond this, meadows roll towards the horizon, their green expanse interrupted only by several cows standing placidly, chewing on grass.

Abruptly the air darkens. A few raindrops fall. Then it turns into a steady downpour. There's a flash of light. Lightning forks through the sky, zigzagging in its search for the ground, and finds the steeple of the church. As it does so, a crashing boom echoes across the landscape. The church shivers as the fiery streak of electricity grazes it, but it holds its own.

The rain now falls even harder, and lightning again splits the sky. Once more it smashes down on the church steeple, accompanied by the percussion of thunder. Then lightning flashes and strikes the church a third, fourth, and fifth time. Each time, the wooden structure shudders, but doesn't collapse.

Suddenly a voice rises above the sound of the storm. It seems to emerge out of the sky itself, broadcasting over the entire village and speaking in an upper-class British accent. 'I say, fellows, don't you think that's somewhat sacrilegious?'

Immediately the rain stops, and light comes back. Its harsh glare reveals a towering array of electrical apparatus surrounding the miniature model village. Metal crossbars loom on either side. A network of wires criss-crosses the sky. Above it hangs a rain machine, still dripping water.

A group of men, five standing and one in a wheelchair, watch the scene from 20 feet away. Giuseppe Faccioli, the man in the wheelchair, is leaning forward, as if eager to see more of the simulated lightning storm. His crippled body looks frail, an impression accentuated by his slightly effete fashion choices: bow tie, hair parted carefully in the middle, thin moustache. However, his eyes, magnified by his thick, round glasses, sparkle with a boyish excitement. He shifts in his chair to look at the man who spoke. 'Mr Walker,' he says with a heavy Italian accent, 'you have a question?'

Walker, an English visitor to the lab, frowns unhappily. 'I'm sorry to have interrupted, Professor Faccioli, but must the lightning always hit the church? It seems somewhat sacrilegious to me.'

Faccioli leans back in his chair, a contemplative look on his face. Finally, he nods. 'The steeple attracts lightning.'

'But can't you make it hit the trees, or the other houses?'

Faccioli shakes his head. 'We do not aim the lightning. But do not worry, Mr Walker. Our lightning arresters will protect your church.' He gestures towards the model, where the small church stands unharmed.

Walker harrumphs. 'It just seems downright peculiar to me that the church is always hit.'

'I agree. Lightning is a peculiar phenomenon, Mr Walker. Very peculiar. That is why we study it. But rest assured, we hold no malice towards the church.'

Walker scowls. 'I see.'

Faccioli continues, 'With your permission, may we proceed with the demonstration?'

Walker frowns again, but after a moment nods his head. 'Yes, of course.'

Faccioli turns his head and nods at an engineer sitting at a control panel. As he does so, he rolls his eyes, ever so slightly. The engineer notices the gesture and begins to laugh, then catches himself and turns away from the group so they can't see his grin. He busies himself adjusting the switches and gauges.

The lights dim again. Rain starts to fall on the village. There's a hum as giant capacitors charge. Then a fork of lightning leaps down and crashes onto the spire of the tiny wooden church.

Tens of thousands of years ago, when lightning flashed through the sky and thunder rumbled across the land, early humans cowered for shelter in caves or hid beneath trees, gazing up with rain-streaked faces at a force that was awe-inspiring and seemed impossibly powerful – a weapon of the gods. But human nature being what it is, it didn't take long for fear and awe to turn into desire and envy. Men dreamed of wielding that fearsome power themselves. They yearned to hurl lightning bolts and make the ground tremble like the gods.

Greek mythology describes one of the earliest attempts to make this fantasy a reality. Salmoneus, King of Elis in southern Greece, was said to have built a brass bridge over which he rode his heavy chariot to simulate the crashing of thunder. As he did so, he hurled lit torches at his subjects, as if casting down lightning bolts. His soldiers ran along behind him and speared to death anyone whom his torches hit. The whole display must have been scarier than real lightning to the terrified citizenry. But it seems the gods weren't amused by Salmoneus's encroachment on their power. They struck him dead with a well-aimed lightning bolt of their own.

Centuries later, the mad Roman emperor Caligula imagined himself a living god. To encourage others to share this belief, he paraded around wearing a golden beard and carrying a brass rod fashioned to look like a lightning bolt. He also tried to create real thunder and lightning. The Roman historian Cassius Dio wrote that Caligula had 'a contrivance by which he gave answering peals when it thundered and sent return flashes when it lightened'. What this contrivance might have been is anyone's guess. The modern historian Steven Scherrer speculates it probably involved 'some sort of rapidly combustible mixture'. However, for all his pretensions to being a lightning god, Caligula didn't fare any better than Salmoneus. His guards stabbed him to death.

Lightning = Electricity!

For almost 1,700 years, no one improved on Caligula's lightning contrivance. Then, in 1708, an English clergyman, William Wall, made a serendipitous discovery that led to new insights into the nature of lightning and ultimately paved the way for scientists to create lightning of their own. The discovery happened while Wall was searching for a way to manufacture phosphorus.

Phosphorus had only recently been discovered, and the substance fascinated researchers because it glowed in the dark. Strong demand for it meant that anyone willing to manufacture it could sell it for a nice profit. However, the only known way to produce it was through a laborious (and smelly) process of boiling and distilling large amounts of urine. Wall was hoping to find an easier and less noxious method of manufacture. Dried faeces, he had already found, also contained phosphorus, but this only exacerbated the gross factor. Hoping to avoid excreta, he started to test other materials to determine if they had phosphoric qualities.

One of the first substances Wall examined was amber. He obtained a long rod of it and, while sitting in a darkened room of his house, rubbed it vigorously with a woollen cloth, 'squeezing it pretty hard with my hand'. As he did so, the amber gave off brilliant little sparks that popped and snapped with a sound like burning charcoal. This wasn't exactly like the glow of phosphorus, but in a way it was even more interesting, because Wall noted that the tiny sparks looked like miniature lightning. What Wall didn't realize was that the sparks were a form of electricity, and that his amber rod represented the first step towards artificial lightning.

Other researchers soon made the connection that Wall missed – that the sparks were electrical – and when they did it seemed logical to conclude that lightning must also be an electrical phenomenon. However, it wasn't until 1752 that an experiment devised by Benjamin Franklin confirmed this hunch.

Franklin proposed that a researcher should raise an iron rod 30 or 40 feet in the air, insulate it at the bottom to prevent current from running into the ground, and then wait for bad weather. The rod would serve as an atmospheric electricity collector. If it became electrified as a storm cloud passed overhead – evidenced by drawing sparks from the rod – this would prove the electrical nature of lightning clouds. The experiment was relatively simple, but also incredibly dangerous. It was like asking to be hit by lightning. Franklin downplayed the danger, but he didn't rush out to stick a pole in a lightning cloud. Instead, it was two French gentlemen, Comte de Buffon and Thomas François Dalibard, who read about his idea and decided to make it happen.

Buffon and Dalibard arranged for the experiment to be conducted in the town of Marly-la-Ville, just outside of Paris. Carefully following Franklin's plan, they erected an iron rod, 40 feet high, that rose out of a sentry box in which a researcher could stand protected from the elements. But like Franklin, they opted not to expose themselves to unnecessary danger. Instead, they found an old, agreeable, and presumably expendable local resident, Monsieur Coiffier, and told him what to do. They patted him on the back, and cheerily said, 'Tell us if it works!' Then they returned to the safety of Paris.

Obediently, day after day, Coiffier sat by the sentry box waiting for bad weather. Finally, on 10 May 1752, his patience was rewarded. Grey clouds rolled in, and he heard a loud clap of thunder. Immediately he rushed into the sentry box and carefully held out an insulated brass wire towards the rod. He heard a crackling noise and a large spark leapt from the rod to the wire. Excitedly, he screamed, 'It worked! It worked!' The prior of Marly heard the screaming and feared something awful had happened. He dropped the book he was reading and sprinted to Coiffier's aid, followed close behind by a crowd of his parishioners. To the prior's great relief, he found the old man unharmed, and the two of them spent the next fifteen minutes enthusiastically drawing off sparks from the pole until the storm passed.

Franklin, to his credit, conducted a version of the experiment soon after, famously substituting a kite for the iron rod. Despite many illustrations that show his kite being struck by lightning, this never happened. Like the French rod, his kite acquired a charge from the presence of atmospheric electricity. Franklin said he detected this charge by extending his knuckles towards a key he had tied to the bottom of the string and receiving a sharp electric shock. He was lucky he didn't suffer more serious injury, but others didn't share his good fortune. In August 1753, a lightning bolt killed Professor Georg Wilhelm Richmann while he was conducting a version of the experiment in St Petersburg, Russia. In his newspaper, *The Pennsylvania Gazette*, Franklin noted that Richmann's death was a tragedy, but added, 'The new doctrine of lightning is, however, confirmed by this unhappy accident.' A few years later, the chemist Joseph Priestley remarked that any scientist should feel lucky to die 'in so glorious a manner'.

The confirmation of the electrical nature of lightning made Franklin famous throughout the world. Praise and awards showered down on him. The Royal Society awarded him their Copley Medal, which was the eighteenth-century equivalent of winning a Nobel Prize. The German philosopher Immanuel Kant even declared Franklin to be a 'Prometheus of the modern age'.

Thunder Houses and Copper Bladders

Now that men knew the secret of the gods' fire, they lost no time in unleashing their own inner lightning gods. They did this by reigning down terror on miniature villages. The miniatures were called 'thunder houses'. They were essentially dollhouses, often made to look like little churches. Add some fake lightning, supplied by the spark from an electrical device such as a Leyden jar, and BOOM! The charge would cause a wooden insert to come flying out as if the house was exploding. Even more dramatic versions included gunpowder to create actual explosions. Miniature people were sometimes added for extra fun. For instance, in 1753, Franklin's friend Ebenezer Kinnersley advertised a new thunder house he had built which featured an artificial flash of lightning 'made to strike a small house, and dart towards a little lady sitting upon a chair, who will, notwithstanding, be preserved from being hurt; whilst the image of a negro standing by and seeming to be further out of danger will be remarkably affected by it.'

Thunder houses allowed researchers to demonstrate the effects of lightning in a sensational fashion, but they also let them show off their newly acquired ability to shield structures from danger, thanks to another one of Franklin's inventions: the lightning rod. One spark exploded a thunder house, but touch a spark to a miniature house protected by a tiny lightning rod and the charge escaped harmlessly to the ground.

The sets and models quickly became more elaborate. In 1772, a London draper and electrical enthusiast, William Henley, thought it would be a nice touch if the lightning came from an actual cloud – or something that looked like one – so he fashioned a fake cloud from a 'bullock's bladder of the largest size', which he obtained from his 'ingenious friend' Mr Coventry. The ingenious Mr Coventry gilded the bladder for him with leaf copper, and then suspended it from a wooden beam. Henley gave the gilded bladder a strong electrical charge, and then brought a brass rod close to it. When he did so, the bladder threw off its electricity 'in a full and strong spark'. It gave new meaning to the phrase 'discharging the contents of your bladder'.

Along similar lines, the Dutch scientist Martin van Marum created artificial clouds out of hydrogen-filled bladders that floated around his laboratory. He charged one cloud positively and another negatively. When they drifted close together, a spark leapt between them. Sometimes, for the amusement of audiences, he placed a third (non-charged) bladder cloud between the two charged ones. When a spark passed through it, it produced a satisfying explosion.

The high point of this model mania arrived in 1777, when the London-based researcher Benjamin Wilson constructed a lightning-creating apparatus that was 155 feet long and hung five feet off the ground, suspended by silk ropes. He housed it in the Pantheon, an Oxford Street dance hall. Because the lightning machine was too big to move, he rolled miniature houses towards it, pushing them closer with a stick until a forked tongue of energy leapt out and smashed into them.

The Surge Generator and the Modern Jove

As elaborate as the eighteenth-century lightning models were, researchers were keenly aware that their simulations were mere firecrackers compared to the mighty roar of the real thing. They simply didn't know how to produce bigger bangs, but gradually, as the nineteenth century passed by, the advance of electrical science addressed this shortcoming. In the 1830s, the first electromagnetic generators were invented, capable of producing far more power than the electrostatic devices of the eighteenth century. In 1882, Thomas Edison, as we've seen, opened the first commercial power plant in New York City. Then, in the 1890s, Nikola Tesla, after severing his relationship with Westinghouse and striking out on his own as an independent inventor, constructed a massive transformer capable of throwing 100-foot-long electrical arcs across his laboratory.

A famous photograph shows Tesla calmly sitting in a chair next to his transformer, nonchalantly reading a book as a fiery storm of electrical sparks dances overhead. Unfortunately, the scene isn't real. It was created using multiple exposures. Being that close to the arcing electricity would have killed him. Also, although Tesla's arcs looked impressive and were high voltage, they were low amperage, and therefore not equivalent to real lightning. A machine capable of producing blasts on a par with lightning still eluded scientists. Mankind's mastery of Nature was incomplete.

The next two decades witnessed the Golden Age of Electricity. Power lines snaked across the countryside, directing current along carefully planned routes into cities and towns. Electrical devices – ovens, lights, and radios – appeared inside homes where they buzzed and hummed reassuringly. The demon of electricity had been tamed and transformed into a domestic servant. The symbolic climax of this achievement occurred on 2 March 1922, when the General Electric Company unveiled the world's first true artificial lightning machine inside its lab in Schenectady, New York.

The machine, whose technical name was a surge (or impulse) generator, towered two storeys high, looking like something pieced together by a mad scientist from spare parts found at a power plant. There were racks of foil-covered glass plates held up by metal crossbars, banks of vacuum tubes, insulators, and other mysterious-looking equipment. Front and centre stood two imposing brass globes mounted on wooden posts. These were the 'sphere gaps' between which the lightning jumped.

Beside the surge generator stood its proud inventor, Charles Proteus Steinmetz. He was a curious-looking man, especially when placed beside such an imposing piece of machinery. He suffered from dwarfism, barely reaching four-and-a-half feet in height. He also had a hunchback and hip dysplasia, causing his torso and legs to bend at an awkward angle to each other. To complete his look, he sported a thick, bushy beard, wore pince-nez glasses, and always had a large cigar clamped in his mouth. Though Steinmetz's body was twisted and crippled, his mind was brilliant. After emigrating to America from Germany in 1889, he became General Electric's most valuable employee when he devised the mathematical equations that allowed engineers to understand the transmission of alternating current. There were rumours that GE didn't even pay him a salary: they just handed him cash whenever he asked for it. The rumours weren't true, but he *was* very well paid. His wealth just added to the contradictions that surrounded him, since he was simultaneously a staunch socialist who once offered his electrical services to Vladimir Lenin.

Inside the Schenectady lab, Steinmetz paraded in front of his lightning machine as reporters scribbled in their notebooks and photographers snapped pictures. He boasted to the audience about his invention's power:

> In our laboratory, we have built a lightning generator that can produce a discharge of ten thousand amperes, at over a hundred thousand volts; that is, a power of over a million horsepower, lasting for a hundred-thousandth of a second. Although this is only one five-hundredth the energy of a natural lightning bolt, it gives us the explosive, tearing and shattering effects of real lightning.

A true showman, he rubbed his hands together as he stepped forward to demonstrate its capabilities. He pulled a lever, and the giant machine emitted a loud buzz as it charged up. The reporters stepped back nervously. Some placed their hands over their ears. Suddenly there was a blinding flash of light, and forked bolts of energy leapt out from the sphere gaps and smashed into a large wood block placed between them. A deafening crash shook the laboratory, and a cloud of dust billowed up. When the dust settled, the reporters could see the block of wood had disappeared – vaporized by the lightning bolt. Chunks of it had landed up to 25 feet away.

Other objects were blown up: a small tree, pieces of wire, and a model of a village. The next day, in the breathless reporting of the event, almost every newspaper hailed Steinmetz as a 'modern Jove'. The *New York Times* even included the epithet in its headline, 'Modern Jove Hurls Lightning at Will'. Benjamin Franklin had only been a Prometheus. Steinmetz had gained the status of Zeus himself. Again, this only added to his contradictions, since he was an outspoken atheist.

Steinmetz died the following year – not from a lightning bolt like King Salmoneus, but due to a heart attack in his sleep. The Italian-born researcher Giuseppe Faccioli continued his research at GE, soon doubling the output of the lightning machine to generate 2,000,000-volt bolts. Like Steinmetz, Faccioli was physically handicapped, but whereas Steinmetz could walk (with difficulty), Faccioli was confined to a wheelchair. A personal attendant pushed him everywhere he needed to go. The *New York Times* thought it a curious coincidence that the two modern masters of lightning were both handicapped. 'As with Steinmetz,' it noted, 'his physical infirmity seems to accentuate rather than diminish the intensity of his mental energy. Both men seem to have taken something vital and tremendous into them from the gigantic forces which they control.'

Despite his handicap, Faccioli was a bit of a daredevil. He liked to get as close to the lightning machine as possible during its operation. He told reporters, 'It is interesting to feel your mustache rise up from the electricity when you are that close.' At GE's Pittsfield laboratory in Massachusetts his engineers constructed an entire model village, and then pounded it with artificial rain and lightning. The lightning whipped back and forth, repeatedly striking a miniature church until, at last, a British visitor stepped forward, believing Faccioli had somehow rigged the church to be hit as a protest against religion. 'But, I say,' the visitor complained, 'don't you think that is somewhat sacrilegious?' None of the surviving accounts of Faccioli mention his religious beliefs, but he had been a close friend of Steinmetz, so it's possible he shared his scepticism. Nevertheless, he hadn't rigged the miniature church in any way. He explained to the visitor that the church spire was repeatedly struck simply because it offered the lightning the quickest path to the ground.

Year after year, the power of man-made lightning grew. By 1929 the Pittsfield laboratory was producing 5,000,000-volt artificial lightning. At the 1939 World's Fair in New York, GE unveiled a machine capable of discharging 10,000,000 volts across a 30-foot gap. It towered 34 feet high, and was housed, as a tribute to its original inventor, in Steinmetz Hall. It proved to be the most popular exhibit at the fair, attracting 7 million visitors who sat on wooden benches and watched the booming and crashing of the energy bolts. Among the visitors was Helen Keller, who, despite being deaf and blind, came away deeply moved. She later wrote, 'It impressed me as nothing else had with a sense of man, frail yet indomitable, mastering the dread smithies from which fiery bolts are hurled. As I sat there, taut but exultant, another wonder befell. The lightning spoke to me! – nay, it sang all through my frame in billowing, organlike tones.'

Electrified Sheep

Descendants of Steinmetz's surge generator can still be found in industrial labs throughout the world. They're regularly used to test anything that needs to be engineered to withstand lightning strikes, such as insulators, aeroplane parts, and church steeples. But in recent decades the generators have also been directed against a more unusual target: sheep.

Sheep and lightning have a long and intertwined history. The animals were (and still are) frequent victims of lightning strikes. In 1939, a single strike in Utah's Raft River Mountains felled 835 sheep that were huddled close together on a hillside. When such scenes occurred thousands of years ago, they must have created a connection in the minds of men between sheep and lightning. Ancient people evidently concluded that the lightning gods had a taste for sheep, because in many cultures sheep became the sacrificial animal of choice to appease lightning deities.

Among the Atuot of southern Sudan, if lightning struck a home, a sheep would be thrown into the burning hut. The Kikuyu of eastern Africa treated people struck by lightning by slaughtering a sheep at the place where the strike happened, and then smearing the afflicted person's body with the contents of the sheep's stomach. Ancient Etruscan priests, the haruspices, held a ceremony of consecration wherever lightning hit: they piled up the objects struck or scattered by the divine fire, and then sacrificed a lamb. Even the Bible mentions the sheep-lightning connection. The Book of Job (1:16) tells us, 'The fire of God is fallen from heaven and hath burned up the sheep.'

It was during the eighteenth century that scientists first took over from their religious counterparts in the business of lightning-related sheep sacrifice. On 12 March 1781, members of the British Royal Society gathered at the home of Rev. Abraham Bennett to witness a sheep exposed to a simulated lightning strike. The men crowded around three sides of a table on which a sheep lay. At the head of the table stood John Read, holding a pair of metallic rods attached to a large battery of Leyden jars that would supply an electrical shock – the faux lightning strike. (The setup was similar to, but scaled up from, the bird experiments discussed earlier.) Read pressed one of the rods tight against the sheep's head, so that its wool wouldn't obstruct the charge. Then he completed the circuit with the other rod. Immediately there was a loud report. Read later wrote, 'Although my situation in so doing [i.e. holding the rods in his hands] was judged dangerous, yet I did not feel the least sensation.' The unfortunate sheep wasn't so lucky. It didn't survive. The experiment demonstrated that scientists could generate enough electricity to kill a large animal, but as a lightning simulation it was useless, since the Leyden jars couldn't muster enough power to come close to replicating even a fraction of the force of the real phenomenon.

The problem of inadequate power plagued all simulated lightning-strike experiments until Steinmetz invented the surge generator in 1922. Once invented, it was inevitable that someone, sooner or later, would decide to stick a living creature inside of one.

As gruesome as such an experiment might sound (and, to be sure, it sounds pretty bad), researchers felt there was compelling medical necessity to justify it. Lightning kills more people every year than any other natural disaster, and yet by the 1920s very little was known about the physiology of lightning injuries – such as how lightning travels through the body, why lightning strikes can be fatal, or, even more mysteriously, why they often aren't. So doctors had scant knowledge to guide their treatment of lightning-strike victims. They couldn't simply apply what they knew about electric shocks to lightning injuries, because it was known, even then, that the two were significantly different. Lightning exposes a body to massive voltages, but only for a very brief amount of time, whereas electric shocks typically involve lower voltages and more prolonged exposure.

The first biological experiments with a surge generator took place in 1931, when Johns Hopkins researchers Orthello Langworthy and William Kouwenhoven investigated whether the route lightning travels through a body makes any difference to its lethality. Rats served as the unlucky victims – mercifully they were etherized, fully unconscious, so they felt nothing.

Langworthy and Kouwenhoven posed the rodents in various positions on the plate of a surge generator, either lying down or hanging upright, and then unleashed a blast of electricity. They discovered that the route of the current made a big difference. When lying down, the animals often survived the experience with little or no signs of damage. It was only when the rats hung vertically, and the shock passed through the entire length of their body from head to tail, that they invariably died. It would be easy to conclude from these results that if you find yourself in an open field during an electrical storm, you should lie down flat, but that would be a mistake. Doing so could expose you to current travelling across the surface of the ground, and you'd end up like the vertically hung rats. The best strategy is to reduce your height by crouching, but touch the ground only with your shoes.

It was another thirty years before lightning researchers turned their attention to sheep. The impetus came in 1961, when a group of insurance companies in the American Midwest approached James Raley Howard, a veterinarian at Iowa State University, with a curious problem. Farmers were filing claims for dead sheep that, so they said, had been struck by lightning, but the insurance companies suspected fraud. Their investigators were pretty sure the sheep had actually died from uninsured conditions such as infectious diseases, but they had no way of proving this. The companies asked Howard if there were any forensic guidelines for identifying lightning injuries and differentiating them from other causes of death. Howard admitted there weren't, so with some financial help from the insurance companies, he set out to create some.

The first stage of Howard's research was non-experimental. For months he remained on twenty-four-hour dead-sheep alert, driving out to remote farms, trying to arrive as soon as possible after a reported strike to examine the animal's injuries. This eventually yielded a checklist of typical signs of lightning injury that included rapid putrefaction, exudation of bloody fluid from the nostrils, singeing of the hair, branched or straight streaks of raised frizzled hair extending from the back to the foot, and lesions in subcutaneous tissue including haemorrhages and dark-brown arboreal patterns. Of course, the surest indicator of death by lightning was a big hole blown in the ground surrounded by dead animals for 50 yards around.

In order to verify that lightning was the cause of the subcutaneous lesions he had noticed, Howard next moved on to experimental research. This involved placing sheep in a custom-built surge generator, housed inside a concrete-block building on the Iowa State campus. Members of the physics department who came by to look at the machine expressed concern that Howard might accidentally blow himself up with the device, but, undaunted, Howard forged ahead.

The sheep stood in a crate between the electrodes of the generator. When Howard activated his machine, 1,000 amperes pulsed between the electrodes, travelling through the sheep. It's hard not to feel sorry for the animals, though Howard noted they were old ewes that would otherwise have been put down by other means anyway – such is the fate of farm animals. At least the end was swift. The process reportedly created 'one hell of a racket', and it invariably destroyed the instruments that monitored the sheep's vital signs. In Howard's words, 'The sheep would die and so would the machine.' But after sixteen sheep, he had collected enough data to verify that lightning does indeed produce the lesions he had noticed in the farm animals. And he managed not to blow himself up.

Howard's research proved to be a good investment for the insurance companies. Their payments for claims related to livestock lightning fatalities dropped by over a third.

These experiments answered the question of what lightning injuries look like, but it wasn't until the 1990s that researchers determined exactly why lightning strikes can be fatal (and why they're often not). Again, it was sheep that helped to provide the answers.

Chris Andrews of Australia's University of Queensland set out to explore the puzzle of lightning fatality using several Barefaced Leicester sheep as his subjects. Like Howard before him, Andrews built his own, custom-designed surge generator. He described it as a 'multiple pulse high voltage impulse generator' that generated six pulses of power, fifteen milliseconds apart. He designed it to mimic the fact that a single lightning flash often consists of multiple strokes of energy, sometimes as many as twenty or thirty.

Andrews' sheep lay fully anaesthetized on a table. Their shaved rumps, wetted with saline solution, rested on a metal plate through which the current could run to ground. Cords wrapped around the animals' stomachs and heads prevented them from moving – and also, presumably, stopped them flying across the room from the force of the discharge. When Andrews turned the generator on, a loud hum filled the room as the capacitors charged, and then, for a split second, a white-hot radiance engulfed the body of the sheep.

Through careful postmortem examination, Andrews discovered that most of the charge from the strike passed over the exterior of the body, in what is known as the 'flashover' effect. But current also entered through open orifices, such as the eyes and mouth, and there it disrupted the functioning of the brainstem, causing cardiac and respiratory failure. However, Andrews concluded that heart failure was not the primary cause of death – although this had long been widely assumed to be the case in lightning injuries – because the heart can restart itself, and often does. Instead, respiratory failure was the greater danger because the lungs can't restart themselves. Andrews advised doctors that even if a lightning strike victim appears to be dead, there's a good chance he or she can be saved if lung function is restored.

Since 1993, researchers have no longer been confined to creating lightning in a lab, because in that year scientists succeeded in triggering lightning strikes in storm clouds by firing rockets into them. The rockets trailed thin strands of wire that connected the clouds to the ground, thereby initiating a strike, and vaporizing the wire in the process. Unfortunately the process was far from perfect. The rocket had to be launched at just the right moment, so that it struck when the cloud was almost ready to release its charge.

But an even more impressive technology is currently under development – laser-triggered lightning. A high-power laser is used to create a plasma channel in the air, which initiates a discharge of energy from clouds. It's like a very sophisticated version of Franklin's idea of sticking a pole into thunderclouds. With such high-tech tools at their disposal, researchers are sure to dream up ever-more ambitious lightning experiments in the future. Sheep beware!

CHAPTER TWO

Nuclear Reactions

During the early decades of the twentieth century, physicists discovered strange forces at work in the subatomic world. They found massive amounts of energy lurking, unexpectedly, in the centre of atoms, and they learned that one element could transform mysteriously into several other elements. At first, these processes seemed illogical and unexplainable. But gradually, after years of patient study, the physicists realized that what had initially seemed random was actually governed by strict rules. Complex mathematical equations could predict, with elegant precision, the behaviour of the subatomic particles. And then, most amazing of all, the physicists discovered they could trick the tiny atoms into releasing the energy hidden within them. When arranged in the correct fashion, atoms would obediently explode in devastating fury. The atomic bomb was born. However, what remained as illogical as ever were the human minds now in possession of the bombs. In fact, the newly discovered atomic power seemed to have a peculiarly intoxicating effect upon scientists. Their behaviour became more erratic, and their ambitions more grandiose, than ever before.

Psychoneurotic Atomic Goats

Bikini Atoll, Marshall Islands – 30 June 1946. The sailor crouches down and scratches the goat on the head. 'Hey, buddy. Time for me to go.' The goat doesn't acknowledge the man's presence. It stares straight ahead, munching its hay with a look of manic intensity in its eyes.

The goat stands on the forward deck of a battleship. Upright metal bars on either side of it restrict its movement; a neck collar secures it to the bars. A bucket of water sits in front of it, and a bale of hay is within easy reach. Several other goats, identically restrained, chomp hay with equal zeal.

The sailor continues. 'Yeah, I know you're busy. Just wanted to say take care of yourself.'

'Hey, Joe!' a voice calls out from the rear of the ship. 'Hurry up!'

'I'm coming!' he yells back.

The sailor stands up. 'So anyway, guys. Enjoy your hay. Hopefully I'll see you soon.' He pats the goat again on its head, and then he sprints towards the stern and disappears from view. Unmoved by his departure, the goats proceed with the business of chewing.

Ten minutes later, a patrol boat pulls away from the side of the battleship. The boat's motor sputters angrily, but as the distance between the boat and the ship widens, the noise grows fainter until at last it can't be heard. The humid tropical air lays a heavy blanket of quiet over the ship. The only sounds are the lapping of waves and the relentless chomp, chomp, chomp of mastication.

A warm breeze blows across the water. Visible over the bow is an entire armada of military ships anchored at varying distances – a few nearby, but most clustered two miles away. Small patrol boats, like tiny specks on the water, shuttle busily between the ships.

Time passes. The sun dips low on the horizon, casting long shadows over the ocean. The small boats depart, abandoning the larger ships, which float silently, swaying with the current. The jaw muscles of the goats flex as their teeth grind hay into pulp.

The sun sets behind a fiery curtain of colour. Stars appear above, twinkling in the equatorial sky. Briefly the goats sleep, kneeling down between the metal bars, but as soon as the sun rises in the east, they wake and resume their work – chewing, chewing.

The sun grows brighter, chasing away the coolness of morning. High *above, a plane flies through the sky. It starts to bank away, speeding towards the horizon. The goats pay it no attention.*

Suddenly there's a brilliant flash of light. Several miles away, over the cluster of ships, a gigantic ball of energy explodes into life. It swells larger and larger, impossibly big, swallowing entire vessels. Then, in the blink of an eye, it transforms into a towering pillar of purple smoke that lifts miles upwards into the sky. The goats ignore it. There's hay to be chewed.

A shock wave races across the water towards the ship, turning the ocean dark beneath it. Within a second, it slams into the vessel. There's a thundering roar, as if the air is splitting apart. The metal of the ship groans from the pressure. A hot wind whips across the deck. Spray and debris fly everywhere. The goats adjust their balance as the ship shudders violently, and then, with steely, unwavering resolve, they lower their heads and take another bite of hay.

The world's first atomic bomb exploded above the New Mexico desert in the early morning of 16 July 1945. Scientists and military officers watched from ten miles away as the blast momentarily illuminated the surrounding mountains brighter than the daytime sun, and a mushroom cloud rose seven miles into the sky.

The second and third bombs detonated less than a month later over the Japanese cities of Hiroshima and Nagasaki. The men, women, and children in the streets, going about their daily business, looked up and saw three American planes in the sky. They thought nothing of it as for months the US Air Force had been sending similar small sorties that usually dropped propaganda leaflets, but on this day the planes had a far deadlier payload. There was an indescribably bright flash. Moments later, the ancient cities were engulfed in flames.

In the weeks and months that followed, people everywhere struggled to understand the implications of the awesome and terrifying power of the atomic bomb. Newspapers filled with urgent questions. What did the existence of the bomb mean for international relations? Would any nation be able to challenge the dominance of the United States? Would America share the secret of atomic power? Was there any possible defence against such a weapon? These were the issues that captured the public's attention, but away from the limelight, in the halls of academia, a few scientists pondered a far more peculiar question. What psychological effect, they wondered, would the bomb have on psychoneurotic goats? On 1 July 1946, when the world's fourth atomic bomb detonated above the Bikini Atoll in the South Pacific, it provided an answer to that question.

Operation Crossroads

The story of the fourth atomic bomb, and of the psychoneurotic goats, began in the offices of the US Navy soon after the surrender of the Japanese government in August 1945. As the admirals and captains viewed the photos of the horrific damage inflicted on Hiroshima and Nagasaki, an uneasy feeling came over them. It wasn't the injuries suffered by Japanese civilians that concerned them. They were worried about their own future. Could any ship, they asked each other with a note of apprehension in their voices, possibly survive against such a weapon? More to the point, had navies (and thus their jobs) suddenly become obsolete?

Motivated by these anxieties, the Navy's top brass devised Operation Crossroads. The idea was to use an entire fleet of ships as target practice for a single atomic bomb and find out just how much damage would be done. If the fleet sank, that would be bad news for the Navy. If much of it survived, that would be good news. They chose the name to signify that the science of warfare stood at a crossroads, and that this test would point the way forward.

As befitting the larger-than-life nature of the atomic bomb, Operation Crossroads was a massive undertaking. The Navy boasted it was the largest experiment ever conducted. The statistics were mindboggling: 42,000 men, 242 ships, 156 aeroplanes, 750 cameras, 5,000 pressure gauges, 25,000 radiation recorders, and 2 atomic bombs – the first to be dropped from a plane and the second detonated underwater three weeks later. All of this military might descended upon the scenic Bikini Atoll in the South Pacific, the site of the experiment. The residents of Bikini, a peaceful people who had managed to survive the war without experiencing any serious conflict, were summarily told they had to move. They were given a vague promise that they'd be able to return when the US government was finished. (They're still waiting.)

Accompanying the officers and sailors were large numbers of researchers, because Operation Crossroads was intended to be a scientific as well as a military experiment. There were nuclear physicists, chemists, mathematicians, spectroscopists, roentgeno-logists, biophysicists, biologists, veterinarians, haematologists, piscatologists, oceanographers, geologists, seismologists, and meteorologists. And accompanying the researchers was a veritable Noah's Ark of animals: 5,000 rats, 200 mice, 60 guinea pigs, 204 goats, and 200 pigs. The animals were to be transferred onto the target ships in order to serve as a stand-in for human crews. Their bodies would reveal both the potential fatality rate of an atomic attack, and the kinds of injuries that might be sustained.

The use of the animals sparked widespread public outcry. Over ninety people wrote letters to the Navy offering to man the target ships in place of the animals. The volunteers were a motley collection of the suicidally inclined who liked the idea of going out with a bang. Many of them were geriatrics who didn't expect to live much longer anyway, but there was also an inmate from San Quentin's death row. One correspondent requested that, in the event the military should desire his services, his name not be disclosed because he didn't want the public to think him crazy. A declassified report noted that, for the purposes of the test, humans actually would be 'more satisfactory than animals'. Nevertheless, the researchers turned down all the human volunteers.

Some of the animals had purposes more specialized than being mere cannon fodder. There were strains of mice bred to have greater or lesser likelihoods of developing cancer, in order to allow biologists to measure the carcinogenic qualities of the bomb. Medical doctors slathered other animals in lotions, cut their fur to simulate human hairstyles, or dressed them in mock military uniforms and 'anti-flash' suits. This was done to test the radiation-shielding properties of different creams, hairstyles, and fabrics. And at the request of psychologists, several goats were brought along that displayed psychoneurotic tendencies.

The psychoneurotic goats were the only animals at Bikini specifically intended for psychological experimentation. As *Bombs at Bikini*, the official report of Operation Crossroads published in 1947, explained rather enigmatically, the goats were there because researchers wanted to find out what effect the 'severe explosion phenomena' would have on their neurotic tendencies. The idea was to place the goats on a ship far enough away from the zeropoint of the drop to ensure they wouldn't be killed by the initial blast, but close enough to get the full effect – to feel the hot maw of the bomb breathing down their necks.

Richard Gerstell, a senior radiologist present at the operation, later included a more succinct description of the goats' purpose in a *Saturday Evening Post* article in which he discussed his experiences at Bikini. Cornell University researchers, he revealed, had handpicked the goats, and they were there 'so that the scientists would be able to see what effects the nuclear explosion had upon their nervous systems, and in this way perhaps the human susceptibility to crack-up and panic'. In other words, it was an experiment in terror. The sensitive creatures were to be forced to gaze up at the screaming atomic behemoth towering up from the ocean, seven miles high, and the scientists would observe their reaction. The researchers wanted to find out whether the vision of atomic Armageddon, impossible for the goats to comprehend, would instantly cause the animals to slip downwards into the murky abyss of madness.

The Soldier's Sensitive Heart

No scientific article ever detailed the reasoning behind the psychoneurotic goat experiment, nor did any reporter ever investigate their presence at Bikini. If it weren't for the brief discussions of the goats in the Official Report and Gerstell's article, we wouldn't even know they existed. They wander out onto the solemn stage of Operation Crossroads, bleat provocatively, and then disappear from sight. For this reason, it's tempting to dismiss them as a one-off, a peculiar study dreamed up by some unknown psychologist for the occasion of the atomic-bomb test. But this isn't the case. There's an entire history of scientific investigation that preceded them, a context that explains not only why researchers were worried the bomb might trigger 'crack-up and panic', but also why they specifically chose goats for the test. To understand this history, it's necessary to jump back a few decades to a time before the Bikini bomb drop, when doctors first noticed a disturbing phenomenon – that the brutal conditions of modern warfare seemed to be causing soldiers, en masse, to go insane.

War has always been hell, but around the mid-nineteenth century, thanks to advances in military technology, it became distinctly more hellish than before. Amid the slaughter of the American Civil War, military doctors began seeing a curious new kind of patient in hospitals. These were soldiers who appeared outwardly healthy, with no obvious injuries, but they clearly weren't well. They exhibited symptoms such as shortness of breath, severe fatigue, and heart palpitations. In fact, the men could barely function. The physician Jacob Mendes Da Costa conducted a study of these soldiers and concluded they suffered from a form of cardiovascular disease. He said they had a 'vulnerable heart'. His colleagues called it 'soldier's heart'.

During World War I, the carnage shifted to Europe. Gunshot, bomb, and poison-gas victims streamed into hospitals, but again, as the Americans had earlier found, there were many soldiers who showed no visible injuries but were nevertheless completely incapacitated. They shivered uncontrollably, vomited up whatever they were fed, and lost control of their bowels. The sheer number of these patients disturbed physicians. These trembling soldiers made up over 10 per cent of all the cases in the wards. Doctors now began to suspect it wasn't merely a heart condition they suffered from. Instead, it was some kind of nervous breakdown or mental paralysis. The terrors of modern warfare – huddling in trenches, helplessly exposed to shell bursts, sirens, and dive-bombing planes – seemed to be more than some minds could endure. War was driving men insane. The doctors came up with a new name for the condition: 'shell-shock'.

It was all well and good to have a name for the condition, but what the generals wanted to know was how to cure the problem. Huge numbers of their troops were unable to fight, for no apparent physical reason. How could the men be patched up and sent back out to the front lines? The doctors had no answer. The best they could do was ship the worst patients back home to rot in mental hospitals, and urge the less extreme cases to pull themselves together. The generals helped out by advising that if anyone thought about deserting, they'd be shot.

Finally the war ended, which seemed to take care of the problem, but then World War II arrived, and with it shell-shock returned with a vengeance. Now it wasn't just soldiers who suffered from battle neurosis. Civilians started to buckle under the stress of constant air-raid sirens and bombing campaigns.

Desperate to do something, military doctors sought more proactive treatments. With the insights of modern scientific psychology to guide them, they were sure they could somehow defeat the invisible enemy of mental terror. They reached into their psychological tool bag and pulled out solutions.

The most obvious remedy would have been to stop the bombing and bloodshed, but that, of course, wasn't an option. Instead, psychologists came up with the idea of 'immunizing' soldiers to the horrors of war before they ever set foot on a battlefield. The theory was that men wouldn't fear what they were accustomed to, and without fear they wouldn't suffer mental collapse. Under the guidance of researchers, the British Army created 'battle schools' in which soldiers trained with live ammunition as the sounds of war played over loudspeakers. The schools contained a core of common sense. It was sensible to expose soldiers to more realistic conditions during combat training. But soon the battle academies led to an even more radical idea – hate training.

The theory of hate training was that burning rage would immunize a soldier to fear even more effectively than mere familiarity, and thus guard him against neurosis. The syndicated columnist Ernie Pyle offered this summary: 'The flight surgeons say that a man will fly better, fight better, survive longer, if he is burning with a violent personal hatred for his foe. For then he will go beyond himself, he will become a temporary fanatic, and he won't worry over his own possible death.'

The British Hate Training Academy opened in April 1942. Participating officers were taken into a 'hate room', where they were shown pictures of enemy atrocities – rotting corpses, starving people, sick and diseased prisoners. They watched sheep being killed in slaughterhouses, and then smeared themselves with the blood of the animals, screaming with rage as they did so. They attacked balloons shaped like men. As they plunged their bayonets in, the balloons burst, showering them with blood. They crawled on their stomachs through mud, past simulated booby traps, as instructors ran beside them shouting, 'On, on, kill, kill . . . Hate, kill, hurt . . . Hate, kill, hurt!'

News of the hate training disturbed the British public. The London *New Statesman* labelled the concept 'wicked bunk', deriding it as 'a series of experiments designed to manufacture the emotions of hatred and blood lust, for all the world as if [the soldiers] were Pavlov's dogs'. Teaching soldiers to hate also didn't seem very British. After all, weren't the British supposed to be the good guys? As it quickly became apparent that the training wasn't having the desired effect anyway – it made soldiers depressed rather than filling them with an urge to fight – the military quietly shuttered the Hate Training Academy.

Artificial Blitzkriegs

Even if the hate training had worked, it was merely preventive medicine. It did nothing for those whose minds were already shattered. In particular, it couldn't help the thousands of men, women, and children who had suffered nervous breakdowns during the Battle of Britain in 1940, when German bombs rained down on British cities. To treat these existing cases, the military psychologists F. L. McLaughlin and W. M. Millar proposed the theory of 'deconditioning'. The idea was similar to the concept of immunizing soldiers to the horrors of war, but it would be immunization after the fact. The doctors hypothesized that they could remove fear through familiarity. They would expose patients, in a safe environment, to the sounds that terrified them – air-raid sirens (or 'Moaning Minnies' as they were called), rifle fire, and exploding bombs. Many patients were so on edge that the mere squeaking of a door sent them into a state of screaming panic, but repeated exposure to the offending sounds, the doctors hoped, would soon desensitize their fear response.

Because they had no audio equipment, McLaughlin and Millar first used a small portable field siren and 'an assortment of tin boxes and sticks' to simulate the sounds of war. These didn't generate much of a reaction from the patients. But then they obtained recordings of actual warfare made by BBC technicians who had placed microphones throughout the country during the German bombing campaign. These recordings proved far more effective. In darkened hospital wards, at midnight, the doctors played the recordings of wailing sirens and rattling gunfire. Patients ran screaming from their rooms, but the doctors persisted. They stood beside the terror-stricken patients, holding their hands and patiently repeating the phrase, 'These are the ones which won't hurt you.' The repetition finally paid off. After a few months, the researchers found that 'small children have been so de-conditioned by the records that they keep on playing with their toys under actual bombardment conditions'. Perhaps the treatment had gone slightly too far.

Although the doctors could reproduce the sounds of warfare, the setting of the hospital ward was unrealistic. McLaughlin and Millar wondered what would happen in more true-to-life surroundings. So they arranged a full-scale, on-location test. In September 1941, they herded a group of bomb-shocked neurotics – young and old, male and female – into an underground London shelter. There they subjected their patients to an 'artificial blitz bombing'. Loudspeakers blasted the sounds of bombs exploding and wailing air-raid sirens. An Associated Press reporter present at the test described the scene: 'The sounds swelled in the dark vault. The guns kept banging. Then big bombs burst. The guns kept up. More bombs. Then the crackle of flames. Next clanging fire engines added their noise, the other sounds continuing.'

The reporter swung a flashlight around the darkened vault and saw tense and anxious faces. However, no one fainted or cried. All remained calm. So the researchers increased the volume. Still there were no nervous breakdowns. The happy researchers declared that the test 'proved their theory that whole populations could be exposed to "artificial blitzkriegs" and thus rendered immune to fear during air raids'.

In the United States, doctors weren't confronted by the problem of bomb-shocked civilians, as their European counterparts were. However, the American military did open a similar 'Battle Noise School' in the South Pacific, led by Commander Uno Helgesson, to rehabilitate battle-broken soldiers. Helgesson placed the nervous men in trenches, dugouts, and foxholes. Then, as described in the *Manual of Emergency Treatment for Acute War Neuroses*, he subjected them to 'mock strafing, land mine explosions and simulated dive bombing attacks'. Once deemed cured, the men were sent back out to fight. Statistics on the number of cures are unfortunately not available.

Experimental Neurosis in Animals

Hate training and deconditioning were ultimately rather desperate, ad hoc attempts to deal with an overwhelming situation, but between the wars researchers had stumbled upon a phenomenon that, they hoped, would eventually yield a better understanding of war neurosis, and better ways to treat it. They discovered it was possible, through experimental means, to create neuroses in animals. In the laboratory, animals could reliably be transformed into nervous wrecks. The ability to do this, researchers anticipated, would allow them to analyse neurosis in a more controlled, systematic fashion. As *Parade* magazine noted in a 1950 article on the subject: 'By establishing mental disorder in animals – less complex than human beings – science now has a simple way of studying such disorders without having its findings thrown awry by complex human emotions.'

The Russian researcher Ivan Pavlov was the first to create an experimental neurosis in an animal. Pavlov was famous for his work on conditioned reflexes in dogs, for which he won a Nobel Prize in 1904. He trained mongrel dogs to associate the ringing of a bell with the arrival of their food. Soon, he discovered, the ringing of a bell caused the dogs to salivate in anticipation, whether or not he gave them any food. Their salivation reflex had come under his control. He had conditioned it.

Around 1917, Pavlov was conducting a variation of his conditioned reflex experiment. He trained a dog to expect meat powder every time it was shown a circle. If Pavlov showed it an ellipse, that meant no meat powder. Pavlov then progressively made the circle and the ellipse similar in shape until eventually the dog couldn't tell them apart. The dog grew confused. It didn't know what to expect. Meat powder or no meat powder? Unable to handle the frustration of not knowing, it suffered a mental breakdown. It barked violently, wriggled about manically, and bit the surrounding equipment. Pavlov wrote, 'It presented all the symptoms of a condition of acute neurosis.'

In America, several of Pavlov's students subsequently expanded on this finding. During the 1930s, William Horsley Gantt, working at Johns Hopkins University, turned his dog Nick into a neurotic basket case. Using an experimental setup similar to Pavlov's, Gantt taught Nick that, in order to receive food, he had to distinguish between two tones produced by a metronome. Once Nick had mastered this skill, Gantt began making the tones increasingly similar in pitch until finally Nick couldn't tell them apart. Like Pavlov's dog, Nick rapidly descended into a state of acute neurosis. At first, he merely exhibited signs of restlessness – whining, panting, and violent shaking. But then he began to refuse, under any conditions, to eat Spratt's Ovals, the brand of food given to him during the experiment. He would only eat Purina Checkers.

What Gantt found interesting was that, as time progressed, Nick's neurotic behaviour grew more pronounced, even though he had been removed from the experimental situation. Nearly three years after the tone training, Nick began fear urinating whenever he was brought to the lab – in the elevator leading up to the lab, in the corridor outside it, and then about once a minute in the lab itself. A year later, the urination progressed into 'abnormal sexual erections' whenever Nick encountered anything that reminded him of the tone-training experience. Even if Gantt himself met Nick away from the lab, such as on a farm, the dog immediately developed a 'prominent erection' and ejaculated. Gantt took full advantage of the dog's condition and often showed off the peculiar trick to visiting colleagues. Gantt merely started the metronome ticking, and Nick immediately developed a terror-induced erection. Gantt enthusiastically wrote, 'We could always count on Nick for a demonstration.'

Another of Pavlov's students, Dr Howard Scott Liddell, became a professor at Cornell University. There, in 1927, he began applying Pavlov's method of creating experimental neuroses to goats. To create a psychoneurotic goat he first attached a wire to a goat's forelimb in order to be able to give it a mild electric shock. Then he used the clicking of a telegraph sounder to warn the animal of the impending shock. Tick, tick, tick, tick, *zap*! When he repeated this process every six minutes, twenty times a day, a goat soon showed signs of agitation, but if he repeated the warning and shock every two minutes, goats quickly froze in a state of 'tonic immobility', every muscle in their body tense and their forelimb, to which the zapper was attached, extended rigidly upwards. Once the experiment was over, the goats limped awkwardly out of the lab, even though nothing was physically wrong with their legs. Back outside, they walked normally. A *Parade* magazine reporter witnessed one of Liddell's goat experiments:

> When the shock comes the animal jumps. Shocks, *without advance warning*, can go on indefinitely without giving him 'nerves'. But when the warning signal is introduced – that's when the trouble starts. At the sound of the bell, or the flashing of the light, he builds up tension or anxiety. When this kind of thing happens regularly he has a breakdown. Like the nervous clerk, who had been having anxieties and shocks in his job, the goat was stumped and confused, couldn't sleep nights. He became neurotic, shy, shunned other goats, became overexcitable.

Here, Liddell speculated, was the secret of neurosis. It wasn't the shock itself that caused the breakdown. It was the nervous anticipation of the shock, enduring hour after hour of tension, knowing another jolt would soon arrive. This insight could be applied directly to battle neuroses. 'It may prove useful,' Liddell wrote, 'to think of the war neuroses as experimental neuroses.' Liddell imagined a connection between the goats in his lab, frozen in fear as they waited for the next electric shock, and people cowering in bomb shelters or soldiers huddled in foxholes waiting for a bomb to burst. In both cases, the monotonous, unyielding stress eventually broke their minds.

In 1937, Cornell gave Liddell 100 acres of land for him to use as an animal behaviour farm. The local press called it the 'heebie jeebie farm'. Like some kind of Orwellian fantasy land, it was populated by the neurotic subjects of Liddell's lab: Achilles the Nervous Pig wandered the property alongside Homer the Neurotic Goat. An Associated Press article about the opening of the farm assured readers that the agitated animals would serve a useful purpose. They would 'imitate man's artificial civilization to discover the cause and cure of nerves, psychoses and delinquency'.

By the mid-1940s, Liddell was boasting of his ability to 'select at random a sheep or goat and confidently predict the type of experimental neurosis that will develop when it is subjected to rigid temporal conditioning'. Liddell also worked with dogs, pigs, and rabbits at the farm, but he preferred sheep and goats. Dogs and pigs, he said, were 'too complex in their behaviour', while the rabbit was 'too simple'. Sheep and goats were just right.

The Psychoneurotic Goats of Bikini

With the scientific context of Liddell's experiments in mind, the logic of bringing psychoneurotic goats to Bikini resolves into focus. Military doctors were deeply concerned about the possible psychological effect of the A-bomb. The reputation of the bomb had rapidly swelled in the year following the bombing of Hiroshima and Nagasaki. Almost supernatural powers were attributed to it. It seemed plausible that the experience of cowering beneath a mushroom cloud might distil into a few moments a mental horror equivalent to weeks in a foxhole. It might trigger instant neurosis. And if the military was going to send soldiers to fight an atomic war, it needed to know how they might behave on the battlefield. So Liddell's psychoneurotic goats were going to provide an answer. They would serve as proxies for human soldiers. Therefore, the US military evidently requested that Dr Liddell, the mad-goat expert, select a few goats for their use. He chose the ones with a skittish look in their eyes, the sensitive ones most likely to react to the sight of an atomic bomb.

After the goats received Liddell's seal of approval, certifying they were fully psychoneurotic, they were transported from Cornell to San Francisco, and from there to Bikini. In the final week of June 1946, sailors loaded them onto a target ship, the USS *Niagara*, anchored two miles from zeropoint. Researchers then set up a protected movie camera and trained it on one of the goats to record its reaction. Finally, they awaited the dropping of the bomb.

On the morning of 1 July 1946, a B-29 Superfortress took off from the island of Kwajalein carrying Gilda, the name given to the Nagasaki-style bomb. People around the world gathered in front of their radios to listen to live coverage of the event. As the wheels of the bomber lifted off the runway, a reporter announced, 'The plane is airborne. The atom bomb is in the air on its way to Bikini, for the greatest experiment, the most explosive experiment in history!'

An hour before the drop, radio audiences began to hear an ominous *tick tock tick tock*. It was the sound of a metronome being broadcast from an unmanned transmitter on board the *Pennsylvania*, one of the target ships. When the ticking stopped, it would mean the bomb had exploded.

At regular intervals, a military voice interrupted the broadcast of the metronome to announce the time remaining before the drop. 'Ten minutes to go, ten minutes to go.' Then, 'Two minutes to go, two minutes to go.' At almost exactly 9 a.m. Bikini time, listeners heard: 'Bomb away, bomb away,' followed almost immediately by: 'Listen world, this is Crossroads!'

The bomb fell through the air. For the first few seconds, it glided along on a path almost parallel to the plane, and then it began to veer downwards, descending at a rate of 300 miles per hour. At 518 feet above the lagoon, it detonated.

A spherical shock wave ripped through the atmosphere. Initially it travelled at a rate of 10,000 miles per hour, but over a distance of three miles its speed reduced to the velocity of a violent gust of wind. Simultaneously a gigantic fireball formed over the water. For a few seconds it shone intensely bright, shimmering with a blue-tinged potency, and then it swept upwards into a massive mushroom cloud that, within five minutes, had reached higher than Mount Everest.

The naval officers nervously waited four hours, and then, brushing aside concerns about radiation, sent in boats to survey the damage. They had to know – how many ships had sunk! Closest to zeropoint, the sailors cruised quietly past burning wreckage and twisted pieces of steel that used to be ships. But the good news, from the Navy's perspective, was that only five ships actually sank, though many more were rendered inoperable. The ships furthest out, such as the *Niagara*, suffered minimal damage. The Navy breathed a sigh of relief. Perhaps the atomic bomb wasn't so terrible after all. Cynics later noted it might have been better for world peace if the bomb actually had sunk the entire fleet.

The blast killed 10 per cent of the animals. Sailors immediately set to work retrieving the survivors and returning them to the laboratory ship, the USS *Burleson*. Men boarded the *Niagara*, released the psychoneurotic goats from their restraints, and hurried the film of the representative goat's behaviour back to be developed. In a darkened room aboard the *Burleson*, researchers gathered to find out what had happened. Had the goat gazed up at the pillar of fire and water and surrendered its mind to insanity? Richard Gerstell described the strange scene viewed by the researchers on the flickering screen: 'When the film was developed later, it showed the goat calmly eating before the detonation. At the moment of explosion there was a mass of flying objects on the screen; and then, this clearing, the goat once more could be seen eating calmly and very much undisturbed.'

The official report of Operation Crossroads offered a similar account: 'Goats are imperturbable animals. . . . The pictures give a clear view of the goat, and show him munching his hay without interruption as the shock wave struck and debris flew all about.'

The Navy's big fireworks, in other words, hadn't impressed the psychoneurotic goats at all. They didn't seem to notice the bomb going off. Who cares about an atomic bomb when there's hay to chew!

It's possible to interpret the goats' non-reaction in a variety of ways. Perhaps goats are, by nature, dull-witted animals that were ill suited for such a test. Or perhaps these particular goats were so tightly wound inside that not even an exploding atomic bomb could make a dent in their neurotic compulsion to eat hay. But to the Navy, their lackadaisical response, coupled with the survival of much of the fleet, seemed like great news. 'No collapse. No nervous breakdown,' Gerstell crowed in his article. He took this to mean that an atomic war wouldn't present any insurmountable psychological problems. If psychoneurotic goats could handle an atomic bomb, then surely so could soldiers and civilians.

In the days following the test, an almost giddy sense of bravado gripped the Navy. The dropping of the bomb had been preceded by such a build-up of expectations that anything less than the heavens opening up and the fires of the apocalypse issuing forth would have been a disappointment. Seeing that the devastation, upon cursory examination, was less than anticipated, naval officials acted in a manner similar to the goats – as if the bomb was actually no big deal. Admiral Blandy surveyed the burning wreckage of the fleet and dismissively declared that he had seen plenty of ships more damaged by kamikaze attacks. Sailors piled back into one of the test ships, despite radiologists warning it was 'hotter than all hell', and brazenly cruised around the lagoon as a show of strength. Taking a cue from the Navy, a reporter for the *Christian Science Monitor* sneered, 'The A-bomb is a distinctly overrated weapon.'

Of course, what the Navy failed to appreciate at the time was that radiation doesn't necessarily kill right away. It can take years, or even decades, to inflict its damage. But the Navy soon received a lesson in the power of radiation from the test animals. Although 90 per cent of them lived through the initial blast, two weeks later a medical officer admitted to an Associated Press reporter that these apparent survivors were now 'dying like flies'.

Were the psychoneurotic goats among those that died of radiation sickness? We simply don't know. In fact, their post-Bikini fate is a mystery. The answer may lie hidden somewhere in the reams of technical documentation about Operation Crossroads housed in Washington, DC's National Archives, but it hasn't yet been found. Most of the surviving animals were sent back to US laboratories for further study. It's quite possible the psychoneurotic goats ended up back at Cornell's animal behaviour farm, where they spent their final days frolicking with Achilles the Nervous Pig.

For the most part, the public ignored the slow deaths of the Bikini animals. People preferred to focus on the happy idea that the bomb wasn't quite as fearful as it had previously seemed. The ultimate symbol of this carefree attitude was the debut of 'the bikini' soon after the test. The irradiated lagoon gave its name to a skimpy, two-piece swimsuit designed by the French engineer Louis Réard, who boasted that any woman wearing one of his creations would provoke reactions similar to those of people viewing an atomic bomb blast. Obviously he hadn't heard about the non-reaction of the goats.

However, not everyone ignored the animals. On 22 July 1946, the San Fernando Valley Goat Society organized a ceremony at Fernangeles Park in North Hollywood to honour the goats that sacrificed their lives at Bikini. Members of the society showed up with their own goats in tow. The original plan was for the ceremony to include the playing of taps on a bugle and the lowering of the American flag, but faced with protests from veterans who didn't feel goats deserved such an honour, the society scaled back the ceremony to a simple moment of silence. As the seconds ticked by, the members of the society stood quietly, their hands on their hearts. The goats paid their respects in their own way. They wandered around the park, munching intently on grass.

How to Survive an Atomic Bomb

Bikini Atoll, Marshall Islands – 2 July 1946. Admiral William Blandy and Navy Secretary James Forrestal stand on the prow of the patrol boat, shielding their eyes against the bright sun as they gaze out at a scene of destruction, the result of the previous day's atomic-bomb test. The vast lagoon surrounding them has become a graveyard of ships – once powerful combat vessels casually twisted and ripped apart by the force of the bomb. Fifty yards directly in front of them floats what remains of the aircraft carrier Independence. *Its flight deck is shredded, as if broken into pieces by a giant mallet. Holes gape in its sides, exposing the buckled steel beams of its interior. Its superstructure is flattened like a pancake.*

'Not as bad as I expected,' Blandy says.

Forrestal nods in agreement. 'If this is the worst the A-bomb can do, then our jobs are secure.'

'Admiral,' a sailor behind them calls out, 'the captain of the tug is on the radio. He says the Sakawa is sinking.'

Blandy spins around. 'Let's see that at once,' he orders. 'Full speed ahead.'

The boat's motor roars into life and the vessel speeds off in a northerly direction. Several minutes later it arrives by the side of the Sakawa, *which is in even worse shape than the* Independence. *The 6,000-ton Japanese vessel is already half submerged, its stern resting on the floor of the lagoon. What remains visible is charred and buckled. A hole in its side is rapidly taking in water.*

'They called her the most hated ship in the Pacific,' Blandy comments.

'The men will be glad to see her sink,' Forrestal adds.

Suddenly the Sakawa *lists violently to port and, with a groan of metal, submerges more rapidly. She goes under as they watch. As the last piece of her bow slips beneath the surface, a large bubble of green foam rises up, like a giant belch of pollution, indicating where she once had been.*

'I think it's safe to say none of the test animals on her survived,' Blandy says.

The radiologist James Nolan nervously approaches the two officers, looking down with concern at the square Geiger counter he holds out before him. 'What's the problem, Mr Nolan?' Blandy asks.

Nolan frowns. 'Sir, the radiation readings here are off the scale.'

Blandy laughs and turns to Forrestal. 'Mr Nolan's counter is so deli-cate that my luminous watch dial will make it go off the scale. But OK, let's humour him and get the hell out of here.'

Blandy shouts an order to the captain of the patrol boat, and the motor again roars into life. The boat makes a sharp turn and speeds off towards the far edge of the lagoon.

After its departure, stillness hangs over the final resting place of the Sakawa. *The sun beats down and gradually the dark green stain in the water dissipates. Soon no sign remains of what lies beneath the surface. Four hours pass, and then the whine of a small boat scudding across the lagoon disrupts the silence. A sailor is standing up in the boat, scanning the horizon.*

'Hey, look over there,' the sailor says to his companion. 'What's that?'

The boat manoeuvres around to pull up alongside a small white object bobbing in the water. The sailor leans over the side to examine it.

'Holy cow! It's a pig!' he shouts. 'And it's alive! Whatcha doing in there, little fella? Hey, Frank, help me get it into the boat.' Both men lean over the side of the vessel, grab on to the animal, and awkwardly haul it in. The pig squeals as they do so.

'I didn't know pigs could swim,' Frank says.

'I guess this one can.'

Frank examines its ear tag. 'Number 311. What ship was it on?'

The other sailor grabs a folder from beneath the instrument panel, flips it open, and runs his finger down a column of numbers. He shakes his head, puzzled. 'This can't be right.'

'What's wrong?'

'It says 311 was on the Sakawa.*'*

'The Sakawa*? Didn't that just sink? I thought none of her animals survived.'*

'Well, according to what's written here, that's the boat she was on.'

Frank cocks his eyebrow sceptically. 'That's impossible.'

The other sailor laughs. 'Maybe this pig is indestructible or something!'

Both men stare down at the pig, as if seeking an answer to the puzzle. The pig looks back up at them and grunts loudly.

Histories of nuclear weapons usually pay particular attention to the physicists who built the bombs, men such as J. Robert Oppenheimer and Edward Teller. Unleashing the power of the atom definitely was an impressive intellectual achievement, but there was another set of researchers who also worked on nuclear-related issues and who faced what was arguably a far greater challenge. In fact, it may have been an impossible challenge. The physicists simply had to figure out how to blow everything up. These other researchers tried to figure out how to make sure something remained alive after the bombs fell. What they produced was a curious science of nuclear survival.

The Indestructible Atomic Pig

Research into how to survive an atomic bomb began soon after the conclusion of World War II. However, in a telling sign of its priorities, the American military's first concern was not to learn how to protect the general population from this terrible new weapon, but rather to assess the vulnerability of its battleships. If atomic war broke out, the generals wanted to make sure the ships remained afloat and kept fighting. Thus was born Operation Crossroads.

As previously discussed, the plan for the operation was to simulate an atomic attack on a fleet by collecting together a huge flotilla of target ships in the South Pacific's Bikini lagoon, populating the ships with animal crews (rats, goats, pigs, mice, and guinea pigs), and then dropping a 23-kiloton bomb on the whole lot.

Within hours of the blast, the Navy sent boats back into the lagoon to survey the scene. As expected, all the ships within a 1,000-yard radius sustained heavy damage, but only five of them actually sank. To the Navy, this was a very positive sign, boosting the spirits of the admirals, who had secretly been fearing the A-bomb would make their entire branch of the military obsolete. However, an event that occurred the next day, 2 July, raised hopes even higher that the atomic bomb was not all it was cracked up to be. Sailors unexpectedly found a pig swimming in the lagoon. An examination of her ear tag, which read number 311, revealed she had been locked in a washroom on the main deck of a Japanese cruiser, the *Sakawa*.

The Navy had obtained the *Sakawa* after the surrender of the Japanese and decided, rather than scrapping her immediately, to include her in Operation Crossroads. Since she was an enemy vessel, the admirals wanted to make sure she sank, so they positioned her less than a quarter-mile from the zeropoint of the blast.

Sure enough, she went down in flames, with her superstructure flattened by the bomb and massive holes ripped in her stern. It seemed impossible that any of the research animals onboard could have survived. Nevertheless, here was a pig, still alive and apparently unharmed.

The men brought the pig back to the floating research ship, the USS *Burleson*, where Navy medic Lieutenant Carl Harris examined her. To verify her identity, Harris dug up photographs of the animals placed onboard the *Sakawa*. All the research animals had been photographed and their positions carefully recorded, down to the smallest rat. There in the pictures, posing on the main deck of the *Sakawa*, Harris found Pig 311. He could tell it was the same pig by the size and markings. She was 50 pounds, six months old, of the Poland China breed, and mostly white with black patches. There was only one possible conclusion, unlikely as it seemed. Somehow Pig 311 had survived the fiery blast, had then escaped from the washroom, abandoned the sinking ship, and floated for hours in the lagoon before being found.

The media made Pig 311 an instant celebrity. Headlines trumpeted her miraculous survival. 'This Little Pig Went Swimming After Blast', declared the *Washington Post*. 'Pig Picked Up in Bikini Lagoon Is Still Alive', announced the *Chicago Tribune*. But despite her daring escape from death's grasp, no one expected her to remain alive for long. After all, she had been exposed to a massive amount of radiation, not only from the blast itself, but also from swimming in the irradiated water. Her days seemed numbered.

Confirming the gloomy prognosis, Pig 311 initially showed signs of radiation sickness. Her blood count dropped considerably, and she appeared irritable and restless. But then the miracle of Pig 311 grew even more miraculous, because after a few weeks she seemed to shrug off the effects of the radiation. She scampered around happily, grunting prodigiously. As far as the Navy biologists could tell, she was in perfect health. She had apparently taken the worst abuse modern science could hurl at her and had casually dismissed it, as if to say, 'Is that all you can do?'

Unable to explain her recovery, the Bikini researchers sent Pig 311 back to the Naval Medical Research Institute in Bethesda, Maryland, where scientists continued to monitor her. Every month they took blood samples, which remained consistently normal. It was only a year later, after she matured into a 300-pound sow and the researchers tried to mate her with other pigs, that they uncovered any sign of possible radiation damage. She turned out to be sterile. This was probably a blessing in disguise. Who knows what kind of mutant piglets she might have borne if she had been able to conceive.

The lesson of Pig 311 may have been that the atomic bomb is like a capricious god of destruction. Sometimes it will spare a life, for no apparent reason. But the lesson that the anxiety-ridden, post-World War II society chose to draw from her example was a far more reassuring, optimistic one. Surely if a little pig could survive an atomic bomb, then anyone could survive! Perhaps the A-bomb wasn't so mighty after all!

The columnist H. I. Phillips hailed Pig 311 as 'a symbol of mind over matter and of pork over both'. *Life* magazine fêted her as 'the little animal that defied the big explosion'. *Collier's* magazine declared her survival to be 'a fable for our time', and transformed her experience at Bikini into a story for children – an inspirational tale parents could read to their kids at bedtime. With not a hint of satire, it described to readers how 'Patty the Atomic Pig' (as *Collier's* named her) was flung out of the Japanese cruiser by the blast, but then took advantage of her sudden freedom to party and hobnob with the other irradiated wildlife of Bikini before finally being rescued:

Patricia went flying over and over through the air. Then, all at once, she hit the water of the lagoon with a great *ker-splash*! Down, down, down she went, nose first, until she hit the coral bottom. Scrambling to regain her balance under water, she bumped smack into a striped tiger fish, who angrily demanded, 'What's going on up there?'

The US government, seeking to maximize Pig 311's propaganda potential, moved her out of the Naval Medical Research Institute in 1949 – by which time she weighed 600 pounds – and installed her in a place of honour at Washington, DC's National Zoo. Every day, crowds of admirers shuffled by her pen, gazing in awe at the amazing indestructible pig. Not all of the public unquestioningly accepted her as a symbol of hope. Some viewed her with suspicion. The director of the zoo admitted to a *Washington Post* reporter that he repeatedly received inquiries from people who were concerned she might still be radioactive and pose a threat to visitors and other animals, but the director assured everyone she was entirely harmless.

Life in the Blast Zone

It was all well and good that battleships and miracle pigs might survive a nuclear war, but after the Soviets detonated an atomic bomb in 1949, the American government finally got around to worrying about preserving its population – and it was definitely the Americans who displayed the greatest enthusiasm for creating a bombproof society. The Europeans only very slowly followed in the footsteps of the Americans in this area, apparently reasoning that they'd be ground zero of any conflict between the superpowers, and therefore stood little chance of remaining alive whatever they did. The Soviets were generally more concerned about saving the lives of their top party leaders than the general populace.

To calm rapidly swelling public fears, the American government created the Federal Civil Defense Administration (FCDA), which mobilized scientific resources to explore how the average person might survive an atomic war. From the perspective of the twenty-first century, the most striking feature of these first efforts was how easy many researchers seemed to think it would be to protect citizens against nuclear assault. In 1950, the radiologist Richard Gerstell (whom we met earlier in the company of the psycho-neurotic goats) authored a small book, *How to Survive an Atomic*

Bomb. In it he cheerily advised readers, 'If you are caught outdoors in a sudden attack, a hat will give you at least some protection from the heat flash.' A US government pamphlet, 'Survival under Atomic Attack', offered similar helpful advice: 'There is one important thing you can do to lessen your chances of injury by blast: Fall flat on your face.'

Much of the research from the early 1950s downplayed, or entirely disregarded, the threat of radioactive fallout. In a 1951 article in the *Journal of Social Hygiene*, Harvard professor Charles Walter Clarke analysed public health problems that might arise following an atomic attack, but he expressed no concern about lingering radiation. Instead, he worried that the bomb blasts might disrupt everyday life to such an extent that people would start having too much sex. He wrote, 'There would develop among many people, especially youths, uprooted and anticipating renewed attacks, the reckless psychologic state often seen following great disasters. Under such conditions, it is to be expected that moral standards would relax and promiscuity would increase.' Clarke urged that plans be drawn up to police all bombed-out areas with 'vigorous repression of prostitution and measures to discourage promiscuity'. Clergymen, he suggested, could be transported into the bomb-devastated areas to lead the effort to keep fornication to a minimum.

A 1954 experiment conducted by two researchers at Washington Missionary College, biology professor Lester Harris and nutrition professor Harriette Hanson, showed a similar nonchalance towards radiation. Harris and Hanson wanted to find out what kind of problems a typical family might face if forced to flee a city following a nuclear attack, but they didn't perceive fallout to be one of the hazards. Instead, the challenge, as they saw it, was how to live off the land with minimal supplies and no shelter. So they came up with Operation Survival. This involved leading a group of volunteers on a three-day bivouac in the woods of Maryland. Participants were only allowed to bring a bedroll and a 12-pound survival kit that included first-aid necessities and food items such as powdered milk, dried eggs, crackers, and peanut butter. Most of the volunteers were students from the college, but to give the test a more realistic feel Harris brought along his wife Marjorie and their three children: six-week-old Charles, twenty-month-old Debby, and three-year-old Jay Jay.

Wailing ambulance sirens signalled the start of the experiment. All the volunteers piled into cars, drove 11 miles out of Washington, DC, and then hiked to a remote location in the forest. They set up their camp, such as it was, and searched the woods for edible weeds and sumac blossoms to boil into a sumac cola. As night fell, everyone gathered around a campfire and sang songs. At 10 p.m. they lay down on their bedrolls to get some sleep. It must have been around that time, as they lay staring up at the stars, that it occurred to the researchers that it would have been smarter to try their experiment in the summer, rather than in November, because after the sun set the temperature rapidly dropped to a frigid 28° Fahrenheit. By 5 a.m., Harris conceded it was far too cold for the children. Fumbling around in the pre-dawn darkness, he hurriedly packed them up, hiked out of the forest, and drove them home. He told the press, 'When they started turning blue, I feared it was time to take them back.' The adults struggled along outside for two more nights, though at the end of the test they complained that the contours of the Maryland countryside were permanently impressed on their backs.

Survival City

Research carried out in 1955 at Yucca Flat, 80 miles north-west of Las Vegas in the Nevada desert, added a dose of grim reality to the knowledge about atomic survival. Just as the Navy had learned about the potential vulnerabilities of their fleet by pitting battleships against an actual atomic bomb, civil defence authorities decided that the only way to gauge the damage cities and towns might sustain during a nuclear war was to construct a typical suburban community and then blow it up. The Atomic Energy Commission, Department of Defense, and FCDA jointly coordinated the test, code-named Operation Cue.

Out in the middle of the desert, at a cost of over $1,000,000, researchers built a faux community consisting of ten fully furnished homes that featured amenities such as basements, medicine cabinets, and pantries and refrigerators stocked with food. The houses stood in two rows, one 4,700 feet from the blast and the other at 10,500 feet, in order to provide data about blast effects at varying distances. There were also six small stores, a radio station with two transmission towers, two power lines, gas lines, and two dozen trucks and trailers. The researchers optimistically named this cluster of structures 'Survival City'. They were apparently in a gloomier mood when they chose the street names: Death Street, Doomsday Drive, and Disaster Lane.

Seventy lifelike fibreglass mannequins populated the homes, posed as if they were engaged in everyday activities such as washing the dishes, sitting around the dining-room table, or sleeping in bed. Collectively the mannequins were known as the Darling Family, after their manufacturer. In some of the houses researchers also built indoor bomb shelters, of concrete and wood-beam construction, hoping to determine if these shelters, which had recently become commercially available for around $500 each, would provide any protection in a blast. In one of the shelters they placed a pair of Dalmatian dogs, in another a group of maze-trained rats.

Mrs Marion Jacoby, a forty-three-year-old wife of a Veterans' Administration physician, wrote to the FCDA volunteering to sit inside an indoor shelter in one of the homes along with the dogs and rats, but the civil defence officials denied her request. When she heard the news, she told the press, 'I'm just sick with disappointment.' Her husband's feelings about her unusual request weren't reported.

On 5 May, a 35-kiloton bomb blasted Survival City. Seven of the ten homes survived, though all lost their windows and were severely shaken. A *Los Angeles Times* reporter allowed to tour the homes afterwards described a gruesome scene:

> Some [mannequins] were burned by the blast, with ugly gray wounds in their flesh-colored bodies. Others were speared by

glass and wood fragments. Still others lay decapitated or with legs and arms torn off by the burst. The youngsters were the worst – babies, school children, teenagers hurled by the shock wave into the macabre positions of violent death, rigid, staring, as they would be in death.

Happily, the sheltered animals fared better. Not only did both the Dalmatians survive, despite the house around them exploding in a 'geyser of destruction', but according to Dr Robert Corsbie, the director of Operation Cue, they were 'wagging their tails when they were recovered about seven hours after the shock'. The maze-trained rats also came through the experience unscathed. In fact, back in the lab the next day, the rats demonstrated they were still able to navigate a maze, prompting Corsbie to suggest that humans might show similar resilience and 'be able to undergo nuclear detonation and recover sufficiently within a few hours to carry on their normal routines'.

Recognizing that survivors of an A-bomb might get hungry and want to eat the food in their homes, the researchers also incorporated a frozen-food test into the Yucca Flat experiment. Soldiers entered the blast zone immediately after the detonation and retrieved food from a freezer in one of the homes 4,700 feet from ground zero, as well as from a freezer buried beneath five inches of soil 1,270 feet from the blast. This food – which included French fries, strawberries, chicken pot pie, cod fish fillets, and orange juice – was then fed to a panel of nine food tasters. After careful consideration, the tasters judged most of the food 'fit to be served in any home or restaurant'. The only poor reviews came from two tasters who complained that the chicken pie and orange juice from the freezer closest to ground zero tasted slightly odd, but not odd enough to make it inedible. H. P. Schmidt, director of the food project, concluded that 'in the event of an atomic attack, foods left in a freezer could be eaten safely if not subjected to a direct hit'.

Although the Yucca Flat experiments produced a few reassuring results, such as the survival of the animals and the supposedly edible food, the researchers had to admit that, overall, the situation looked pretty bad. The twisted remains of the Darling Family proved that unsheltered humans couldn't expect to survive anywhere near a nuclear blast. Advice such as 'wear a hat' or 'fall flat on your face' wasn't going to be much help.

Moving Underground

By 1955, scientists were becoming far more aware of the danger of nuclear fallout. A 1954 detonation of a hydrogen bomb at Bikini Atoll (where tests had continued to be conducted after Operation Crossroads) shocked both scientists and the public when it spread a cloud of nuclear radiation over a 7,000-mile swathe of the ocean. Twenty-three Japanese fishermen on a boat 40 miles from the blast spent months in the hospital suffering from acute radiation sickness, and the radio operator on the boat died. Geneticists such as Hermann Muller, who examined the effects of X-ray radiation on fruit flies, were sounding the alarm about the ability of radiation to damage genetic material and produce cancers. And Pig 311 herself might have offered some warning about the long-term hazards of radiation, if anyone had been paying closer attention. She died at the National Zoo on 8 July, 1950, at the young age of four-and-a-half. However, zoo officials didn't record the cause of her death, and the press never reported on her untimely demise.

Acknowledging the lingering threat posed by radiation, survival researchers concluded that if people were going to have any chance of living through a nuclear conflict, they would have to be protected in shelters – preferably not just during the blast itself, but for days or even weeks afterwards until the worst of the fallout dispersed. So during the latter half of the 1950s, the focus of research shifted from above-ground studies to investigations of life underground in shelters. This raised an entirely new concern. Assuming a shelter withstood the initial blast, would the people inside be able to cope psychologically with the stress of living in a confined space for weeks on end?

Shelter experiments began as early as 1955, though the first ones were promotional stunts sponsored by companies trying to sell shelters to a nervous public. For instance, in June 1955 the Houston Home Builders Company sealed the Christmas family (mother, father, son, and daughter) for three days inside a cramped shelter made of steel plates and reinforced concrete. Similarly, in July 1959 Melvin and Maria Mininson spent their honeymoon locked in an 8-by-14-foot backyard shelter in order to demonstrate the habitability of the product of Bomb Shelters, Inc.

The first long-term, scientifically monitored shelter experiment was Project Hideaway, conducted by Princeton psychologist Jack Vernon. His subjects were an average American family – high-school teacher Thomas Powner, his wife Madge, and their three small children.

The study began on 31 July 1959, when the Powners took up residence in a windowless 9-by-8-foot room in the basement of the Princeton psychology building. The Powners lived there for the next two weeks in primitive conditions. They had no electric power, flush toilet, running water, or contact with the outside world. Candles and flashlights provided illumination, and the family heated food in a chafing dish. For enduring this confinement, the Powners received $500.

What the Powners didn't know, until they finally emerged, was that a team of researchers was monitoring and recording their every action via a hidden microphone. Vernon neglected to inform them of this, fearing it would make them self-conscious if they knew. He later admitted, 'They were mad when they found out we were listening, but not as mad as we thought they would be.'

The Powners passed their time reading books and playing games such as Scrabble. Whiskey helped relieve the stress. Madge frequently brooded – so much so that the listening team of psychologists at times feared they were going to have to end the experiment early – but she would loosen up when Tom gave her a drink. Similarly, the middle child, three-year-old Tory, became moody and withdrawn and reverted to wetting his bed, but a tranquilizing shot of whiskey soon set him right.

The biggest problem was the odour. To go to the bathroom the family used a chemical toilet discreetly hidden behind a curtain. However, the toilet didn't decompose or destroy waste materials in any way, so a rank smell quickly built up. This was only slightly alleviated by a hand-cranked ventilation machine in the door. The parents quickly learned to use the ventilator as a reward device. If the kids behaved themselves they were allowed to crank it for a while and get some fresh air.

The Powners mistakenly kept dumping the waste from the toilet in a refuse container intended for their trash. This not only exacerbated the odour, but also triggered a chemical reaction in the container, which started to produce an unnerving moist, bubbling, belching sound. Methane gas leaked from the container into the room, causing both adults to experience dizzy spells.

To boost her morale, Madge put on lipstick every day. This made her feel slightly more normal, despite the filth and stench of the surroundings, but when she discovered their captivity would last a day longer than she had anticipated, she admitted she felt 'shattered'. Nevertheless, upon finally emerging from the room, both Tom and Madge praised the experiment as 'very beneficial'. Madge told the press that the family had gotten to know each other better than ever and that she would willingly do it again.

Shelter experiments flourished during the following decade. Thousands of people spent days or weeks in confined spaces, under the watchful eyes of researchers. In August 1963, researchers even locked up two men with thirty-five cows for two weeks in order to investigate how livestock would react to life in a shelter. After all, people would presumably still want milk and hamburgers after a nuclear war, so cows would need to be protected too!

Most people (and animals) generally experienced few serious difficulties during these confinement experiments. The biggest problems were boredom, insomnia, and bad odours, although there was one case of a man participating in a Pittsburgh-based study in 1960 who became convinced that the researchers were secretly irradiating him through the one-way observation mirrors. He had to be removed early from the shelter.

Psychologists learned that the reactions displayed by the Powners were quite typical. Subjects usually began the experiments buoyed by a sense of adventure and excitement, but by the fourth day this gave way to depression and withdrawal. As the end of the study neared, people grew irritable and prone to outbursts, but upon leaving the shelter, all tension disappeared. Subjects would almost always praise the solidarity of the group and comment on the interesting nature of the experience.

However, one problem troubled the researchers. No matter how true-to-life they made the shelters, they couldn't simulate the sense of danger that would be present during a real emergency. Participants knew the experience would eventually end and normal life would resume. Researchers acknowledged that this limited the applicability of their studies, so some tried to think of creative ways around the issue. For instance, Donald T. Hanifan, a researcher contracted by the Office of Civil Defense to study sources of stress in shelters, suggested in 1963 that hypnosis could be used to convince subjects that a real attack had taken place, or was about to. Letting his imagination run wild, he elaborated on the intriguing possibilities hypnosis might allow:

> For example, a female subject could be told under hypnosis, 'The alarm is sounding, and as you, your husband and son are rushing to the shelter, your husband and son are separated from you in the rushing crowd, and you arrive at the shelter alone.' An additional stress could be imposed by suggesting that she sees her hopelessly inaccessible son trampled on by the panic-stricken crowd. The number of possibilities for varying stresses are too great to enumerate here. If hypnotic suggestion could be used, a wide range of highly informative experiments could readily be designed.

Hanifan acknowledged there were moral problems with this method as well as practical ones, noting it was 'unlikely that suggestions will remain equally effective for all subjects for a period as long as several days'. But he argued that in a time of crisis such considerations might become secondary to the need for knowledge. Thankfully, it never came to that. Or, at least, no researcher ever publicly admitted to following through on Hanifan's suggestion.

Shelter experiments reached a peak of popularity in the early 1960s, and then gradually fell out of fashion, though it took a while for them to disappear completely. As late as the 1980s, a few researchers could still be found locking subjects in underground rooms to observe their reactions.

A number of factors led to their decline. There was the matter of money. Governments woke up to the fact that even if shelters worked it would be prohibitively expensive to build enough of them to protect even a fraction of the population. The easing of Cold War tensions lessened the urgency of planning for nuclear survival. People also grew horrified at the issues raised by survival planning. For instance, in the early 1950s, the Milwaukee school district debated whether schoolchildren should be tattooed in order to make it easier to identify the dead and wounded after a conflict, but decided against it because of the impermanence of tattoos on burned flesh. And in the early 1960s, the so-called 'Gun Thy Neighbor' controversy flared up in the press, in which religious leaders debated whether it was morally permissible for a Christian to shoot his neighbours if they tried to get into his backyard bomb shelter during an emergency. Faced with such dilemmas, more and more people decided they'd rather take their chances and not plan for survival at all.

But the most compelling reason why shelter experiments, and survival research in general, went out of fashion was due to the growing conviction that living through a nuclear war was an exercise in futility. Even if someone survived the initial blasts by huddling far underground in a shelter, then lived for weeks on a diet of canned food and crackers, they couldn't stay down there forever. Eventually they would have to climb out of the shelter, and they would emerge into a radiation-poisoned land. Whereupon they would discover they had only delayed death, not averted it.

Helen Caldicott, president of the group Physicians for Social Responsibility, nicely summed up the legacy of survival research when she was asked in 1982 why government officials and civil defence researchers had continued to believe for so long that it might be possible to survive a nuclear conflict. She bluntly replied, 'Because they're mad. In my experience, they are uniformly ignorant about the medical, scientific, and ecological consequences of a nuclear war.'

Although science did eventually abandon the effort to create a bombproof society, many of the physical artifacts of that effort remain. Out in the Nevada desert, the wreckage of Survival City still stands, serving no purpose now except as a tourist attraction for Las Vegas sightseers. And beneath many backyards, or burrowed into hillsides, bomb shelters still wait to protect someone. Their owners have since found other uses for them. They serve as repositories for old bicycles, tools, and appliances. In recent years, it's become particularly popular to convert them into wine cellars. This is perhaps the best possible use for them. Humanity may not be able to survive a nuclear holocaust, but if the unthinkable ever does occur, at least the last few people remaining alive will have an excellent vintage on hand with which to toast the end.

Nuke the Moon!

Kailua, Hawaii – 8 July 1962. A bright quarter moon peeks through the clouds that drift across the sky. Down below, on the rooftop patio of the hotel, a party is underway. Tourists in bright flower-print shirts chat, drinks in hand. Tanned college boys dressed in T-shirts and shorts lounge at the bar drinking beer. At the end of the bar, a lava lamp lazily burps up bubbles of neon yellow and green.

A young boy who had been staring over the side of the patio at the dark ocean turns, runs over to his parents, and tugs on his mother's shirt. 'Mom, Dad. How much longer until it happens?'

His father pats him on the head. 'We don't know, scout. Hopefully soon.'

The mother glances at her watch. 'If it doesn't happen soon, we're going to have to send you to bed. It's almost eleven o'clock! Way past your bedtime.'

'But, Mom! You promised.'

'Hey, everybody!' a man sitting in the corner by a radio calls out. 'Listen up! They're saying the rocket launched.'

'Turn it up!' someone else shouts.

The man adjusts the volume of the radio, and the tinny sound of an announcer's voice rises above the noise of the party. The crowd falls silent as everyone turns to listen. The voice begins a countdown. 'Ten minutes to go. Ten minutes to go.'

The announcement sends a hush over the party. People mill around anxiously, glancing every now and then up at the sky, speaking in lowered voices. Suddenly one of the college boys shouts, 'The end of the world is nigh!'

'Shut up!' an older man, standing with his wife, angrily rebukes him.

'Maybe it'll be a dud!' another college boy calls out, and a murmur of nervous laughter ripples through the crowd.

'Five minutes to go. Five minutes to go,' the radio announces.

The crowd drifts towards the edge of the patio to gaze out over the ocean. They can see the lights of several ships on the horizon, and they hear the sound of the surf crashing on the beach below. Turning to either side, they see the main street of Kailua, faintly illuminated by the small puddles of light that form beneath each street light.

'One minute to go. One minute to go.'

The young boy grips his mother's hand. 'Mommy, I'm scared.'

'Don't be,' she comforts him. 'There's no danger at all. The scientists know what they're doing.'

The radio voice starts a countdown. 'Ten, nine, eight, seven.' As the numbers approach zero, the voice becomes higher in pitch, as if the *announcer is growing giddy with excitement. No one talks. They all stare expectantly up at the dark sky. 'Five, four, three, two, one.'*

There's a gigantic flash. It's as if, for a split second, the sun materialized in the sky, lighting up the entire landscape bright as day, and then disappeared again. No sound accompanies the flash, but at the same instant, the voice on the radio abruptly cuts off, replaced by the crackle of static. Simultaneously the lights in the street flicker and go out.

Up above, strange things start happening in the sky. The initial white flash spawns a writhing, twisting green stain that spreads rapidly from horizon to horizon, like toxic chemicals pouring into the atmosphere from space. The unnatural colour boils and glows, illuminating the landscape with a vivid radiance that makes the ocean appear a shade of purple, like grape juice.

The crowd watches in silence. Tendrils of the green mass squirm and wriggle, and then they mutate into a hot, angry shade of pink before deepening into dark blood red. The moon, throbbing a violent shade of yellow, comes out from behind a cloud.

An elderly woman cries out, 'It's awful! It's awful.' She clasps her hands over her eyes. Her husband holds her by the shoulders, and moves her towards the doors to go inside.

The young boy's eyes widen with fear. 'Mommy, is the world ending?'

His mother glances down at him with concern. 'Oh, honey. I don't think so. I'm sure everything will be OK.' But as she looks back up at the phantasmagoric colour show twisting luridly in the sky, her eyes betray a flicker of doubt.

By the late 1950s, the United States and Soviet Union, in their desperate bid to outdo each other, had dropped nuclear bombs on all kinds of things. They had levelled fake cities, Pacific islands, naval fleets, and quite a few desert landscapes. As a result, the public had grown complacent, accustomed to a steady stream of press releases announcing new nuclear tests. The generals and military planners were beginning to wonder, 'Is there anything else we can bomb?' Or rather, 'Is there anything we can bomb that will still get people's attention?' The answer appeared above them at night, hovering in the sky like a giant target. It seemed so obvious. Nuke the moon!

Flash Powder on the Moon

Exploding a bomb on the moon was actually an old idea – as old as rocket travel itself. Robert Goddard, the American pioneer of modern rocketry, first proposed it in 1919, though, of course, he suggested exploding a conventional bomb, not a nuclear one.

Goddard was a quiet man, prone to ill health, more comfortable with books than people, but as a teenager his imagination had been captured by the concept of space travel. As he later told the story, at the age of seventeen, he was climbing a cherry tree in his backyard when he looked up at the sky and imagined what it would be like to ascend higher still into the silent void of space. He pictured drifting between the planets, far from the pushy crowds of people on earth. He never forgot that vision. He spent the rest of his life trying to make it a reality.

In 1916, at the age of thirty-four, Goddard was working as a physics teacher at Clark University, in central Massachusetts. In his spare time he designed rockets. The first rockets he built didn't get very high off the ground, but already he had grand plans. He envisioned shooting a rocket all the way to the moon, but he immediately recognized a problem. He would have no way of knowing if his rocket reached the moon. That's when the bomb idea occurred to him. If his rocket carried a bomb that detonated when it hit the lunar surface, the explosion might be visible to someone on earth watching through a large telescope.

To determine how big his bomb would have to be to create a visible explosion, Goddard conducted experiments in the Massachusetts countryside. In the dead of night, away from the illumination of houses and street lights, he tested what the maximum distance was at which he could see a small, explosive burst made by magnesium flash powder. Sitting alone, peering into the darkness, he waited for the tiny pinprick of light to appear in the distance. He imagined he was looking up into space, and that the flash was the first signal from an extraterrestrial source announcing, 'Success! Success!' Using this method, he determined that one-twentieth of a grain of flash powder produced light visible at two-and-a-quarter miles. From this he calculated that if his rocket carried 13.8 pounds of flash powder, it would create an explosion 'strikingly visible' to a telescopic observer on earth.

Goddard described his research and flash-powder experiment in December 1919 in a short treatise titled 'A Method of Reaching Extreme Altitudes'. The publication generated enormous public interest, especially the idea of bombing the moon. He received letters from volunteers eager to be strapped aboard his rocket – would-be lunar suicide bombers. However, many critics derisively dismissed his ideas. The *New York Times*, for instance, published an editorial in which it condescendingly informed Goddard that even schoolboys knew rockets couldn't work in the vacuum of space, since there was nothing there for them to push against. Forty-nine years later, the day after the launch of Apollo 11, the *Times* published an apology, noting, 'Further investigation and experimentation have confirmed the findings of Isaac Newton in the 17th century, and it is now definitely established that a rocket can function in a vacuum as well as in an atmosphere. The *Times* regrets the error.' Unfortunately, Goddard didn't live long enough to see the apology, nor the launch of rockets to the moon. He died in 1945.

Project Cow

Scientific advances made Goddard's lunar flash-powder experiment unnecessary. Researchers learned they could use radio waves to track rockets through space. However, the lunar bomb remained lurking in the imagination of the space community. The idea of looking up and seeing the moon flashing a signal appealed to the gee-whiz sensibility of many rocketeers. The invention of the atomic bomb allowed more firepower to be added to the experiment, so space-age visionaries naturally switched out Goddard's conventional bomb for a nuclear one. But it was the development of long-range rockets in the mid-1950s that really brought the idea to the forefront of scientific thought again. It suddenly seemed possible to reach the moon with a rocket, so why not try lobbing a nuclear bomb up there as well?

Several researchers independently suggested the idea. In March 1957, Dr Harold Urey of the University of Chicago proposed in *The Observatory* that an intercontinental ballistic missile could deliver an atomic bomb to the moon. This would be a useful experiment, he argued, because the resulting explosion might dislodge moon rocks that would eventually rain down on the earth where they could be picked up and analysed.

Three months later, rocketeer Krafft Ehricke and cosmologist George Gamow, writing in *Scientific American*, took Urey's idea a step further. They proposed launching two missiles at the moon. The first would drop an atomic bomb, and the second would fly through the massive cloud created by the explosion, 'sweep up some of the debris blasted from the moon's surface', and bring it back to earth. Because the second missile would fly over and around the moon, they suggested naming the mission Project Cow, after the nursery rhyme. ('Hey diddle diddle/ The cat and the fiddle/ The cow jumped over the moon.') They ended their article on a rousing note: 'Something deep in the human spirit will be stirred when man succeeds in making a cow that jumps over the moon.' And it would be even more stirring, they might have added, when that cow nukes the moon.

The Day the Soviets Almost Nuked the Moon

While such proposals stoked the imagination, they were still entirely theoretical, since no country had yet managed to send a missile into space. That changed on 4 October 1957, when the world learned the Soviet Union had placed a satellite in orbit, Sputnik 1. Less than a month later, on 3 November, the Soviets topped that achievement by successfully launching Sputnik 2, which carried the first living creature into space, a small dog named Laika.

To the United States, these launches were like a wet smack across the face. American leaders believed their nation's global influence depended upon an international recognition of its technological superiority, but now that superiority had been yanked out from under them. Suddenly they were in second place. The Americans feverishly speculated about what the Soviets might do next, and one ominous possibility suggested itself above all others. The Soviets were going to nuke the moon! And the Americans thought they knew exactly when it was going to happen – on 7 November 1957, the fortieth anniversary of the Bolshevik Revolution. Coincidentally, a lunar eclipse was going to occur on that day, which would allow a lunar explosion to be far more visible. It would be the most impressive fireworks display in all of history, on a stage for the entire world to see.

As 7 November approached, near hysterical reports appeared in the media about the Soviet plan to nuke the moon. 'A Soviet rocket might already be en route to strike the moon with a hydrogen bomb,' warned the *New York Times* on 5 November. Soviet United Nations delegate Arkady Sobolev fuelled speculation by letting slip that Russia did indeed hope to commemorate the Bolshevik Revolution with a rocket to the moon. Dr Fred Whipple, director of the Smithsonian Astrophysical Observatory, reported he had heard rumours the rocket was already on the way. The US government appeared impotent in the face of the threat. The best President Eisenhower could do was schedule a 'chins up' television and radio speech to coincide with the anticipated lunar fireworks.

The eclipse was going to be visible for half an hour over much of the Pacific, in an area extending from the Mariana Islands south of Japan to Hawaii. When 7 November arrived, and the earth's shadow slid over the moon, American astronomers trained their telescopes on the lunar surface, nervously looking for the flash of an atomic explosion. They waited and waited, but it never happened. The moon had been spared.

Project A119

The Soviet lunar nuke proved to be a non-event, but the realization that it could have happened, and might still happen, catalysed the United States into action. In May 1958, the US Air Force commissioned a top-secret scientific study, code-named Project A119. For administrative purposes, it was given the innocuous title 'A Study of Lunar Research Flights'. Its true mission, however, was to determine the visibility and effects of a nuclear explosion on the moon.

The study was based at the Armour Research Foundation in Chicago. Leonard Reiffel, head of physics at the Foundation, led the research group. Assisting him were a team of top scientists, including the astronomer Gerard Kuiper and a twenty-three-year-old graduate student named Carl Sagan, who would later achieve fame as the host of the science TV series *Cosmos*.

The Air Force didn't give the researchers any details about when a lunar bomb drop might occur. It told them it wanted the detonation to be a surprise. It also gave them few details about the size of the bomb, suggesting only that it would be about the size of the one dropped on Hiroshima. Based on that information, the Air Force wanted to know where the best place to drop the bomb would be, to ensure maximum visibility from the earth. It also wanted the researchers to come up with scientific justifications for such an experiment. Everyone knew the detonation would primarily be a PR stunt, a way to one-up the Soviets, but it nevertheless had to be wrapped in a veneer of science.

Armed with slide rules and calculators, the scientists got to work. Sagan was given the job of modelling how a cloud of gas and dust would expand above the lunar surface. This was a prelude to determining the visibility of the explosion from the earth. Sitting at his desk, he imagined the detonation – a missile descending out of the blackness of space towards the grey lunar surface, followed by a blinding flash of light. Because the moon has almost no atmosphere, the explosion wouldn't form the characteristic mushroom cloud of a nuclear blast on earth. Instead, the force would radiate in all directions at once. Huge amounts of dust would be blown up into space. Sagan reasoned that this dust cloud would be most visible if it formed at the edge of the moon, where the rays of the sun would illuminate it to dramatic effect. He also suggested that spectroscopic analysis of the dust cloud could reveal valuable information about the composition of the lunar surface, including the possible presence of organic materials.

The researchers studied the issue of bombing the moon for almost a year before issuing their final report in June 1959. It was a curious document. The scientists were obviously uncomfortable with the idea of nuking the moon. The report began by acknowledging that it was beyond the scope of the study to consider the political wisdom of such an action, but nevertheless noted it might generate 'considerable negative reaction' around the world. But with this word of caution behind them, the researchers proceeded to do their duty by assessing the various kinds of scientific information such an event might yield. It was relatively slim pickings. In addition to the spectroscopic analysis already mentioned, the report suggested the blast might produce interesting seismic data as well as add to knowledge about the spread of radiation in space. All of this information, and more, has been gained in the decades since then in less violent ways.

Thankfully, Project A119 never progressed beyond the stage of a scientific study. The Air Force shelved the project after concluding there was too much risk involved without enough reward. However, the Air Force might have changed its mind had it known the Soviets were simultaneously engaged in a similar project that progressed quite a bit further.

Despite American fears, the Soviets hadn't planned to nuke the moon on the fortieth anniversary of the Bolshevik Revolution. At the time, such a feat was beyond their capability. But in early 1958, at the urging of nuclear physicist Jakov Borisovich Seldovich, they initiated Project E-4 to investigate the feasibility of the idea. From this effort eventually emerged a full-scale mock-up of a lunar bomb. It was an evil-looking creation resembling a spherical mine, with initiator rods sticking out on all sides that were designed to detonate the bomb upon impact with the lunar surface.

However, the Soviets, like the Americans, ultimately backed down from the idea of nuking the moon. Their main concern was that the rocket might fail fully to lift the bomb out of earth's orbit – rockets weren't exactly reliable technology. Then they'd have a fully armed nuclear bomb slowly spiralling back down to a random spot on earth, which might cause awkward political issues. The Soviets also concluded that, even if their bomb did make it to the moon, the explosion simply wouldn't be impressive enough. People on earth, if they were looking at exactly the right moment, would see a small bright flash, and then nothing. It hardly seemed worth the effort.

Nukes in Space

The moon's surface remained pristinely nuke-free. But this left military planners with a problem. They still wanted to nuke *something* new, dramatic, and exciting. What else was there?

Edward Teller, the famous creator of the hydrogen bomb, had a few ideas. He'd been an enthusiastic proponent of bombing the moon, but as an alternative he proposed nuking the Strait of Gibraltar. A few well-placed hydrogen bombs, he argued, would seal the Strait shut, thereby causing the entire Mediterranean to fill up like a bathtub. This might mean the unfortunate loss of a few coastal cities such as Venice, but on the other hand it would irrigate the Sahara, so it seemed like an even trade-off. As tempting as the suggestion was, the idea went nowhere, but then Teller came up with an even better plan that became his passion, which he earnestly promoted for years: nuke Alaska!

From Teller's point of view, Alaska had many similarities with the moon. It was far away, seemed pretty barren, and was relatively uninhabited. His idea was that five or six hydrogen bombs could create an artificial harbour on the north-west coast of Alaska, near Cape Thompson. It would be an awesome demonstration of the power of the bomb. Unfortunately, Alaska's residents, few as they were, weren't so thrilled with the idea. They pointed out that a new harbour wouldn't be very useful since the Alaskan coastline there was icebound for nine months out of the year. Teller didn't let logic like that stand in his way. He didn't understand why his critics couldn't see that, useful or not, the instant harbour would be extremely impressive. For four years, from 1958 to 1962, he battled to create his nuclear harbour, before his critics finally managed to sink the idea.

After ruling out the Strait of Gibraltar and Alaska, the Defense Department did, however, eventually come up with a winning target. Casting its eyes upwards again, it started contemplating the idea of nuking space itself – in particular, a specific part of space known as the Van Allen belts.

The Van Allen belts are ghostly belts of radiation surrounding the earth, vast bands of electrons and protons deposited by the solar wind that extend out thousands of miles, held in place by the earth's magnetic field. The physicist James Van Allen, after whom they're named, discovered their existence in early 1958 by analysing the Geiger counter data from America's first successful satellite launch, Explorer 1. When he announced their discovery, some publications struggled to come up with metaphors to describe them. The *Science News-Letter*, for instance, characterized the belts as 'space donuts' because its writers imagined them as giant doughnuts of radiation wrapped around the earth.

'Nuke the space donuts' doesn't have quite the same ring to it as 'nuke the moon', but from the military's perspective, the Van Allen belts offered significant advantages over the moon as a target. First of all, they were a lot easier to reach. Second, while lunar fireworks might look disappointing from earth, a nuclear bomb in the Van Allen belts promised to create a light show without equal. Highly charged particles would rain down on the earth's atmosphere, creating vivid auroral displays stretching over thousands of miles. That was sure to get people's attention.

Also, the Defense Department suspected the blast would have

intriguing military applications. Nicholas Christofilos, a Greek elevator mechanic turned nuclear physicist, predicted the explosion would create an artificial radiation belt around the earth, which would not only interfere with communication systems but also act as a shield, frying the electronics of any missile that passed through it. It could render intercontinental ballistic missiles obsolete. Christofilos imagined creating a permanent shield by detonating thousands of bombs in space every year.

Tempted by the idea of a space-based missile shield, the Defense Department decided to experimentally detonate a nuclear bomb in the belts. Military officials approached Van Allen the day after he publicly announced the discovery of the belts to find out if he'd cooperate in planning such an experiment. He agreed. Historian James Fleming has noted it was an apparently unique instance in which a scientist 'discovered something and immediately decided to blow it up'.

A series of initial, top-secret tests over the South Atlantic, using one-kiloton bombs, took place in August and September of 1958. As Christofilos predicted, the blasts created artificial belts that wrapped around the earth, causing widespread radio and radar interference, but these small-yield bombs were merely a warm-up for the full-scale test, using a 1.4-megaton hydrogen bomb, that the military planned in the Pacific for the summer of 1962.

The military made no attempt to keep the 1962 test a secret. The bomb's effects were going to be far too visible for that. In fact, it was going to be the most widely witnessed experiment in history, producing visible changes in the sky across the Pacific. Anticipating a spectacular light show, the media nicknamed it the 'Rainbow Bomb'.

However, scientists around the world expressed alarm. Dr Martin Ryle of Cambridge worried that such a large explosion might permanently bend the Van Allen belts, or wipe them out entirely, with unknown consequences. Cosmic radiation could come pouring down through the atmosphere, frying huge swathes of the earth. It could trigger a global catastrophe. Protestors marched outside of

American embassies in cities around the world. The ninety-year-old philosopher Bertrand Russell described the planned test as, 'An act of wanton recklessness and a deliberate insult to all men and nations who retain a vestige of sanity.'

To calm these fears, the Defense Department issued a press release assuring everyone that it had analysed the situation, found nothing to be concerned about, and was sure the belts would be back to normal 'within a few days or a few weeks'.

Elaborate preparations were made. Johnston Island, 800 miles west of Hawaii, was chosen as the site of the launch. The Navy transported electronic gear to remote islands throughout the Pacific to record the bomb's effects. On Hawaii, hotel owners advertised rooftop rainbow bomb parties as the anticipated date of the launch drew near.

At a few minutes before 11 p.m. on 8 July 1962, a Thor missile carrying the bomb lifted into the sky. The launch was code-named Starfish Prime. At an altitude of 248 miles, the device detonated.

Huge amounts of high-energy particles flew in all directions. On Hawaii, people first saw a brilliant white flash that burned through the clouds. There was no sound, just the light. Then, as the particles descended into the atmosphere, glowing streaks of green and red light appeared. *Life* magazine correspondent Thomas Thompson reported, 'The blue-black tropical night suddenly turned into a hot lime green. It was brighter than noon. The green changed into a lemonade pink and finally, terribly, blood red. It was as if someone had poured a bucket of blood on the sky.' The man-made aurora lasted seven minutes before fading away.

The lights in the sky weren't the strangest effect of the test. An electromagnetic pulse sent energy rippling through Hawaii's power grid. Power lines fused. Burglar alarms went off. Garage doors mysteriously opened and closed of their own accord. Telephone lines went down. TVs and radios malfunctioned. Hundreds of street lights across the islands flickered out simultaneously.

The military finally had got its attention-grabbing explosion, but it learned an important lesson in the process. H-bombs and satellites don't mix. The electromagnetic pulse crippled seven of the twenty-one satellites in orbit at the time. The casualty list included Telstar 1, the first commercial communications satellite, which, in an accident of bad timing, was launched by AT&T the day after Starfish Prime, causing the satellite to plow directly into the curtain of radiation created by the blast. The blast also temporarily tilted the earth's magnetic field. For thirty minutes the magnetic field in Hawaii was off by a third of a degree. Despite Defense Department assurances to the contrary, it took several years before radiation levels in many areas of the Van Allen belts returned to normal.

The US government realized it was going to have to choose between testing nuclear bombs in space and allowing the development of satellite technology. It chose satellites, and the next year signed the Partial Test Ban Treaty with the Soviet Union, banning all atmospheric, underwater, and space-based nuclear tests. As a result, the world has never again seen the eerie glow of an H-bomb explosion in space. If we ever do see that sight again, it would probably coincide with the collapse of the world economy by crippling many of the hundreds of satellites that now orbit above us, providing us with crucial services such as television, telephone, navigation, and weather forecasting.

Military strategists periodically warn that if any rogue nation or terrorist group wanted to make a statement, the best way to do it would be to launch a nuclear bomb into space. It wouldn't kill anyone, but by knocking out the satellites, it would plunge the world into chaos. Unfortunately, there's little that can be done to defend against such a scenario because hardening all satellites against nuclear attack would be prohibitively expensive.

Nuke-the-Moon Fervour

Although the rainbow bomb made a spectacular show in the sky, lunar bomb enthusiasts were left dissatisfied. After all, the bomb missed the moon by over 235,000 miles. As a result, the idea of sending a nuclear device to our celestial neighbour has remained lurking deep in the collective unconscious ever since, like an unscratched itch, and it bubbles up periodically in fits of nuke-the-moon fervour.

In the autumn of 1990, Alexander Abian, a mathematics professor at Iowa State University, distributed an article to his class in which he argued the benefits of blowing up the moon. He claimed that the pull of the moon's gravity causes the earth to tilt on its axis by 23 degrees. Consequently, the sun heats the earth's surface unevenly. Remove the moon, he suggested, and the earth's weather would improve dramatically, ushering in an era of 'eternal spring'. This could be achieved, he proposed, by a 'controlled total elimination' of the moon using nuclear weapons.

His article was meant as a provocative thought experiment, a way to encourage his students to think big about planetary redesign, but a young woman in his class wrote about his idea in the student newspaper, the *Iowa State Daily*. From there it reached the newswires, sparking an international media frenzy. Abian's plan to nuke the moon appeared in newspapers throughout the world. A German television station sent a team to interview him. A London tabloid ran the front-page headline 'Scientists Plot to Blow Up the Moon'. The *Wall Street Journal* ran a long article analysing the suggestion. Clearly the idea had resonated with the public.

NASA scientists vainly protested the idea. They pointed out that while it's true the earth does tilt on its axis, it's unlikely this is caused by the moon's gravity. The more likely culprit is the accumulation of ice at the poles. More ominously, any attempt to destroy the moon would inevitably cause large lunar fragments to plummet to earth, potentially wiping out all human life. Environmental groups also noted, with concern, that removal of the moon would disrupt the mating cycle of the palolo worms of the South Pacific, since these delicate creatures only rise from the ground to procreate during full moons.

In 2002, nuke-the-moon mania surfaced again, rising to popularity on the Internet after a website, imao.us, posted an article proposing it as a practical means to achieve world peace. The basic argument was that if America started lobbing nuclear bombs at the moon, the rest of the world would be cowed into meek obedience to its demands: 'All the other countries would exclaim, "Holy @$#%! They are nuking the moon! America has gone insane! I better go eat at McDonald's before they think I don't like them."'

To this day, imao.us's 'Nuke the Moon' article remains the most popular page on its site.

Lunar bomb enthusiasts finally had a small bone thrown to them in 2009 when NASA announced it was not only going to crash LCROSS (the Lunar Crater Observation and Sensing Satellite) into the moon, but also televise the event. It was part of an effort to determine if pockets of frozen water were hidden deep in lunar craters. Upon arriving at the moon, the satellite would separate into two sections. The first section would slam into a crater on the moon's south pole, hopefully throwing up a huge cloud of dust. NASA suggested that the debris plume would be visible through telescopes from the earth. The second section would follow behind and fly through the dust cloud, collecting and analysing samples, before crashing into the lunar surface itself. It was essentially a non-nuclear version of Ehricke and Gamos's Project Cow, proposed in 1957.

On the night of the crash, a large crowd gathered outside NASA's Ames Research Center in Silicon Valley, lured by the promise of a lunar explosion. A large screen was set up outside, on which was shown the live feed from the sample-collecting rocket. The same feed was also broadcast on NASA television. The crowd waited with bated breath as the craggy surface of the moon came into view. The NASA announcer counted down the seconds to the impact. The moment arrived. All eyes were fixed on the screen . . . and there was nothing. No debris plume was visible on the camera, nor did earth-based telescopes see anything. NASA later insisted the rocket really had impacted as planned, but explained that the plume couldn't be seen in the visible spectra. Once again, lunar explosion enthusiasts were left disappointed.

Currently (and thankfully) the prospects don't look good for ever seeing a nuclear explosion on the moon. Though you never know. Perhaps one evening you'll be sitting outside admiring the view, gazing up at the tranquil face of the moon, when suddenly you'll see a strange bright flash on its surface. It would be an appropriate moment to repeat the immortal lines spoken by Charlton Heston's character at the end of *Planet of the Apes*: 'We finally did it! You maniacs! You blew it up! Damn you! God damn you all to hell!'

The Incredible Atomic Spaceship

La Jolla, California – June 1958. 'Whoosh! Bleep bleep bleep! Whoosh!' *The five-year-old boy plays with his toy spaceship, a cone-shaped tin rocket with fire-engine-red stripes down the side and matching red fins. He glides it up and down over his head in wide parabolic arcs. 'Whoosh! Whoosh!'*

Through the open window comes the sound of a car pulling into the driveway. Instantly the boy drops the toy and shouts, 'Dad's home!' He runs outside and over to the two-tone turquoise and white Chevrolet Bel Air. His father, a slender man with dark hair, sits behind the wheel with the window rolled down.

'Dad! Dad!' the boy shouts out, unable to contain his excitement. 'Is it true? Are you building a spaceship?'

His father looks down at the boy, a bemused expression on his face. 'So you heard the news?'

'Is it true? Is it true?'

His father smiles and says gently, 'Yes, George. It's true.'

His mother walks out and stands behind him. 'Hello, honey. He's been talking about it all day!'

'Where are you going to fly to? What does it look like? How big is it?' The questions pour out of the boy's mouth.

His father glances at his watch. 'I tell you what, George. Why don't you hop in the car, and I'll show you something you might find interesting.'

George immediately runs around the car, pulls open the door, and leaps onto the seat next to his father.

'Are you going to show me the spaceship?'

'You'll see when we get there.' The man looks at his wife. 'We'll just be gone a little while.'

'Have fun,' she replies.

The father eases the car out of the driveway and onto the main road. He drives for a couple of blocks down palm-tree-lined streets, and then turns up a steep winding road. The Pacific Ocean sparkles in the late afternoon sun to their left.

As they drive, the boy continues to pester his father with questions. 'How fast will it go? Can you take me too?'

His father glances over at him with a conspiratorial expression. 'It's top secret. I'm not allowed to tell you anything, but I'm going to show you just this one thing.'

The boy nods eagerly to show he understands.

They reach the top of a hill and drive for another minute. Then the father pulls over to the side of the road where it borders the top of a canyon. 'Here we are!' He gets out, followed by his son close behind him, and they stand together, gazing out at the view.

Rolling coastal scrub stretches out before them, shadows lengthening as the sun lowers in the sky. His father points towards the edge of the canyon, about half a mile away. 'Look over there.' The boy follows his father's finger and sees something that looks for all the world like a spaceship parked in the middle of the landscape. It's an enormous circular object, about 150 feet wide and 30 feet high, surrounded by thick steel buttresses. Windows ringing the structure brightly reflect the sun.

The boy gasps. 'Is that it? Is that the spaceship?'

His father kneels down beside him. 'Well, no. Not quite. That's the building where I work, where we're designing the spaceship. But you can get an idea of what the ship will look like from that building. Imagine that as the base. Now the ship itself, shaped like a giant egg, will sit on top of that, over 200 feet high.'

The boy gazes down, his mouth wide open, imagining the massive ship rising off the ground and lifting majestically into the sky.

'Where are you going to go?' the boy asks.

His father smiles. 'To the stars, George. We're going to the stars.'

Before the twentieth century, visionaries dreamed up a variety of fanciful methods of achieving space flight. A seventeenth-century bishop, Francis Godwin of Herefordshire, imagined being carried to the moon on a chariot pulled by twenty-five swans. The nineteenth century produced a bumper crop of ingenious schemes. Jules Verne suggested shooting a ship up into space out of a 900-foot cannon. Edgar Allan Poe described a trip to the moon in a hot-air balloon. Edward Everett Hale envisioned placing a brick ship in orbit by rolling it down a giant groove in a mountainside and then giving it a final push with two water-powered flywheels. George Tucker pictured floating into space, buoyed up by Lunarium, an imaginary anti-gravity metal.

By the 1920s, thanks to the work of Robert Goddard, it had become clear that chemical rockets offered the most practical means of launching an object into space, but inventors continued to dream up other techniques. Technological advances simply made their visions ever more extravagant. And so it was that in the mid-1940s the most fanciful scheme of all emerged, unrivalled for sheer audacity, outrageous enough to make even Godwin's swans look sane and reasonable by comparison. The idea was to use an atomic bomb to fling a ship into space, and beyond.

Surfing into Space

The New Mexico desert, in July 1945, provided the setting for the birth of the atomic spaceship concept. The scientists of the Manhattan Project had gathered to witness the Trinity nuclear test, the first explosion of an atomic weapon. As the massive fireball rose above the White Sands Proving Ground, the Project's scientific director, J. Robert Oppenheimer, famously thought of a line from Hindu scripture: 'Now I am become death, the destroyer of worlds.' But his colleague, the Polish-American mathematician Stanislaw Ulam, had far less gloomy thoughts. As he watched the mushroom cloud form in the sky, he wondered whether the massive amount of energy released by an atomic bomb could somehow be used for peaceful purposes. Could it, he pondered, propel a spaceship?

The great limitation of chemical rockets is that the amount of energy you can extract from the fuel is fairly low, and if you try to extract more energy by burning the fuel hotter, the rocket itself melts. Ulam saw a way around this problem. He envisioned getting rid of chemical fuel and the shell of the rocket entirely and replacing them with an atomic bomb, which releases millions of times more energy than its equivalent weight in rocket fuel. Then he would explode the bomb behind the ship. If the ship was designed correctly, with a large steel plate to catch the force of the blast – picture a plate 120 feet in diameter and weighing 1,000 tons – the explosion might not destroy the ship. Instead, it would thrust it forward with enormous acceleration. What Ulam imagined was similar to a surfer riding a wave forward onto a beach, except the wave would be a blast of atomic energy, and it would carry a ship upwards and then through space.

The first time people hear Ulam's concept, they usually think it's crazy. Wouldn't an atomic bomb simply blow up the ship? Even if it didn't, what about radiation exposure, or steering the ship? For anyone but a nuclear physicist, such concerns might have prevented the idea from ever being anything more than a fantastic speculation, but Ulam couldn't get the concept out of his mind. It lingered there, nagging at him, until finally, in 1955, he wrote a report exploring the concept in more detail: 'On a Method of Propulsion of Projectiles by Means of External Nuclear Explosions'.

Ulam's report caught the attention of Ted Taylor, one of America's leading nuclear weapons designers who in 1956 had just started work at General Atomic, a San Diego-based company hoping to capitalize on peaceful applications of nuclear power. Taylor did some back-of-the-envelope calculations and realized, strange as it seemed, that Ulam's idea might just work. He asked his friend Freeman Dyson, widely regarded as one of the most brilliant mathematicians in the world, for his opinion. Dyson was sceptical, but he also crunched some numbers, and, to his own surprise, found himself agreeing with Ulam and Taylor. An atomic-bomb-powered spaceship might actually fly.

Lew Allen's Balls and Manhole Covers in Space

Several events that occurred during the nuclear tests of the 1950s accidentally provided real-life confirmation of Ulam's atomic propulsion concept. In 1955, the Air Force physicist Lew Allen hung large steel spheres (jokingly referred to by his colleagues as 'Lew Allen's balls') near several nuclear explosions, hoping to test the effect of the blast on various materials contained inside the spheres. Allen expected that the force of the blast would fling away the spheres, but they ended up being propelled over vast distances, far greater than he had anticipated. In fact, he had trouble finding them. It demonstrated the propulsive effect of a nuclear explosion that Ulam had predicted.

A second event offered even more dramatic confirmation. During the summer of 1957, the US military conducted a series of nuclear tests, including several underground detonations, at the Nevada Test Site located north-west of Las Vegas. During one of the underground tests, the force of the explosion blasted a massive steel plate – four feet in diameter, four inches thick, and weighing almost a ton – off the top of a bomb shaft. The engineers, realizing in advance this might occur, had trained a high-speed camera on the cover in an attempt to calculate the exact velocity of its ascent, but it moved so quickly that it only appeared in a single frame of the film before disappearing. It was never found.

Dr Robert Brownlee, the experimental designer of the test, nevertheless estimated that the force of the explosion, combined with the focusing effect of the tunnel, had accelerated the steel plate to a speed over six times the velocity needed to escape earth's gravity. In fact, the plate would have had enough acceleration to leave the solar system entirely, soaring aloft like a miniature version of an atomic-bomb-propelled spaceship.

The manhole cover in space, as it soon became known, is one

of the great legends of the atomic age. Its fans like to point out that although Sputnik, the 184-pound sphere launched into orbit by the Soviets on 4 October 1957, is usually regarded as the first man-made object to reach space, the manhole cover actually beat it there by over a month. Unfortunately, the reality is that the manhole cover probably didn't make it into space. It definitely had enough initial acceleration, but the atmosphere would have rapidly slowed it down or burnt it up entirely, scattering it as tiny droplets of metal over much of Nevada. Still, it's fun to imagine that the manhole cover did make it out of earth's orbit and is out there even now, somewhere past Pluto, continuing on its journey as humanity's first, and most unusual, ambassador to the stars.

Project Orion

Lew Allen's balls, and the errant manhole, offered material evidence that an atomic bomb might be able to propel an object into space, but it was Sputnik that provided the opportunity for Ulam's concept to progress from fanciful idea to real-world project. The small Soviet satellite, spinning in orbit through the skies, sent the United States military into a panic. Instantly it became receptive to ideas for getting into space fast – even seemingly far-fetched ideas.

Ted Taylor, sensing the time was right, convinced General Atomic to submit a proposal to the Defense Department's Advanced Research Projects Agency (ARPA), seeking funding for an atomic bomb spaceship. On 30 June 1958, he got the good news. ARPA had awarded General Atomic a million-dollar contract for a 'feasibility study of a nuclear bomb propelled space vehicle'. It looked like the atomic bomb spaceship was going to become a reality.

Taylor quickly assembled a group of top engineers and researchers, including Freeman Dyson. It wasn't hard to recruit talent. The project sold itself. What nuclear engineer wouldn't want to help build a spaceship to zip around the solar system? It appealed to their sense of wonder and fantasy, as if they were a bunch of kids playing with a gigantic, very powerful toy. Taylor named the operation Project Orion. The name didn't mean anything in particular. He just thought it sounded cool.

George Dyson, the son of Freeman Dyson, later wrote a history of Project Orion in which he recalled the day in June 1958 when he first learned his dad was helping to build a spaceship. He was five years old, and he thought it was the most amazing thing in the world. As soon as his father arrived home, he pestered him with questions: 'How big will the ship be?' 'Where are you going to go?' 'Can you take me along?' The engineers working on the project shared young George's enthusiasm.

At the time, other space engineers were thinking small. It required huge amounts of fuel to lift anything into space, so the weight of everything had to be kept as low as possible. For this reason, the first manned space capsules used in the Mercury program were tiny, scarcely more than shells wrapped around the astronauts. The Orion engineers, however, thought big from the very beginning. For them, bigger was better, because it was easier for a gigantic ship to withstand the pounding of the atomic bombs. So their initial estimate, which they considered conservative, was to build a 4,000-ton, 20-storey-high ship – about as big as a nuclear submarine. A 1,000-ton circular disc, the 'pusher-plate', would absorb the energy of the explosions while protecting the occupants of the ship through a shock-absorbing mechanism. They calculated it would require 100 atomic bombs, detonated approximately one half-second apart, to lift this behemoth into orbit.

While NASA engineers thought the moon was an ambitious target, Project Orion set its sights far beyond that. The moon would merely be the first stop on a grand tour of the solar system. The intrepid adventurers envisioned cruising over to Venus. With an atomic-powered spaceship the journey would only take a month. Then they would head over to Mars before blasting their way to Saturn and finally returning home. 'Saturn by 1970' became the rallying cry of the project.

In the idyllic, tropical setting of La Jolla, the San Diego suburb where General Atomic had its headquarters, a kind of manic optimism gripped the Orion engineers. They began to imagine grandiose possibilities. Freeman Dyson worked out that there was no reason the weight of the ship couldn't go as high as 8 million tons, making it 160 times the size of the *Titanic*. A Super Orion of that size could transport a colony of 2,000 settlers on a 150-year journey to Alpha Centauri. Dyson also realized that the power of the Orion spaceship would make massive terraforming projects possible. He envisioned a 'Project Deluge', which involved shipping vast quantities of water from Enceladus (one of Saturn's moons) to Mars, in order to transform our planetary neighbour into a paradise fit for human habitation.

Lift-off at Point Loma

Before letting their imaginations run completely wild, the Orion engineers realized they needed to test the concept of a bomb-propelled ship to make sure it worked in practice as well as in theory. This required building a prototype. Obviously they couldn't use atomic bombs, so they relied on conventional explosives.

Testing took place on Point Loma, a peninsula that wraps around the northern edge of San Diego Bay. The Orion engineers built a one-metre-high, 300-pound model they called 'Hot Rod'. They rigged it up to eject small canisters of C-4, a variety of plastic explosive, through the plate on its bottom. When encased in its fibreglass shell, it looked like a very wide bullet. Without the shell, it resembled an ocean buoy – a small tower on top of a round platform. They spent the summer of 1959 fiddling around, trying to get it to work, and on 14 November 1959, they were ready for a final test.

Jerry Astl, the explosives expert on the project, filmed what happened. There was a loud bang as the C-4 detonated beneath the contraption, and it lurched upwards. A second canister dropped downwards and – BANG – another explosion. It lurched higher again. BANG. BANG. BANG. With each explosion, the Hot Rod looked like it should list over and topple back to the ground, yet improbably it kept staggering higher. Six explosions propelled it up 185 feet. A parachute then deployed, and it landed unharmed. It was a perfect test flight. The concept had worked exactly as the engineers planned.

Reportedly when the review board at ARPA saw the movie, they were speechless. Many of them had doubted the feasibility of Orion, yet here was proof that, at least in theory, the concept worked. With renewed confidence, the engineers got to work on designing the full-scale ship.

Perhaps the most amazing thing about the Orion spaceship was that the engineers believed it could be built with technology that already existed in 1958. They knew how to build bombs, and they knew how to build giant structures such as aircraft carriers. Orion simply married the two forms of know-how.

But, of course, the design of the ship still presented enormous technical difficulties. After all, it was one thing to get a one-metre model to work with C-4; to build a 4,000-ton ship that ran on atomic bombs was something else altogether. For a start, there was the issue of shock absorbers. If you didn't have shock absorbers between the pusher-plate and the ship, the sudden thrust from the bombs would kill the crew, which wouldn't be good. But how do you build shock absorbers to survive multiple atomic blasts?

Second, what about ablation – the wearing away of the steel pusher-plate? How many times could the plate endure being struck by the blast before it disintegrated?

Third, how exactly were you going to shoot the bombs behind the ship? During peak acceleration a bomb would have to go off about every half-second. One could either eject the bombs through a trapdoor in the pusher-plate, and pray the door never jammed, or somersault the bombs around the side of the plate, and hope they didn't fly off in the wrong direction. Neither option seemed ideal. However, these were just engineering problems. The Orion team was sure that, with enough time, they could all be solved.

The Death of Orion

Ultimately it was politics, not technical difficulties, that doomed Orion. The US government formed NASA in 1958, with the expectation that it would assume control of all non-military space projects. However, NASA didn't want to be involved with Orion in any way. The idea of their astronauts sitting on top of a payload of nuclear weapons didn't appeal to the agency. Also, as the NASA administrators noted in a report, 'The question of political approval for ever using such a device seems to weigh heavily in the balance against it.' Therefore, in order to continue to receive funding, Orion had to become a military project. For a while that's what happened. In 1960, ARPA transferred control of Orion to the Air Force. However, the Air Force had no interest in doing grand scenic tours of the solar system. Instead, it required a military use for the project. Thus was born Battleship Orion.

Battleship Orion would have been a truly terrifying creation. Imagine the Death Star parked in orbit around the earth, packing enough firepower to single-handedly make the planet uninhabitable. The ship, protected by its 120-foot steel plate, would have been almost immune from attack. If it detected missiles being fired at it, it could simply accelerate away at atomic speed and hide behind the moon.

The Air Force built an eight-foot model to demonstrate the Battleship Orion concept. It bristled with weapons – 500 minuteman-style warheads and multiple five-inch guns – and also sported various auxiliary landing craft. In 1962, General Thomas Power proudly showed off this model to President Kennedy during a tour of Vandenberg Air Force Base. Kennedy reportedly recoiled in horror, suspecting the top military brass had gone insane. At the time, he was trying to ease tensions with the Soviet Union. Unveiling Battleship Orion would have had the opposite effect. So the model was quietly packed away and has never been seen since. It's probably hidden in a government warehouse somewhere.

In 1963, the United States and Soviet Union signed the Partial

Test Ban Treaty, banning all atmospheric nuclear tests. By making it politically impossible ever to launch the ship, the treaty spelled the end for Project Orion. In January 1965, the Air Force cancelled the project's funding.

To this day, Project Orion remains one of the great what-might-have-beens of the Space Age. As talk of a mission to Mars heats up, the Orion die-hards – and there are quite a few of them – point out that an Orion spaceship could get a crew there comfortably in about one-tenth the time that chemical rockets would take. But of course, there's the issue of radioactive fallout. This might not be so much of a problem for the astronauts, shielded behind the pusher-plate, as it would be for everyone else on earth living with the after-effects of the one hundred explosions it would require for Orion to reach orbit. Dyson calculated that a single Orion launch would produce enough fallout to kill approximately ten people. That, he reasoned, made the technology an impossible sell.

Would Orion actually work? Most of the engineers employed on the project, many of whom are still alive, continue to insist it would. Given this, it's tempting to think Project Orion might not have been mothballed entirely. Maybe the military secreted the project away to Area 51, where they continued to develop it. Maybe a completed Orion spaceship exists today, hidden underground beneath the Nevada desert or in a remote region of Siberia.

Unfortunately, the only time its existence would ever be revealed to the public would be in the case of a mass-extinction event, such as a nuclear war or large-scale asteroid impact. The politicians and top military brass would file on board. Then it would blast upwards, bearing its passengers to the safety of space. It would be a spectacular sight, but it would be one of the last sights all of us non-VIPs ever saw.

Deceptive Ways

Astronomers build giant telescopes to gaze at the stars. Physicists construct billion-dollar supercolliders to unlock the secrets of the subatomic world. Biologists use electron microscopes to view the structures of the cell. By contrast, psychologists seem to have it easy. They don't need any special equipment to see their subjects. People are right in front of them. But they do have a problem other scientists don't. Unlike stars, atoms, and cells, people lie. You can't trust them! For instance, someone asks how you feel. Perhaps you say, 'Great!' though you really feel lousy. It's a little white lie. People tell them all the time. Such untruths facilitate social interaction, but they play havoc with psychological research. To counter these deceptive tendencies, psychologists have developed misleading strategies of their own, on the theory that their subjects are less likely to lie if they don't know they're being studied – or what they're really being studied for. It's like fighting fire with fire. This has created a situation unique in science – the researcher recast as prankster, creating elaborate ruses to pull the wool over the eyes of unsuspecting dupes.

Men Fight for Science's Sake

Berlin, Germany – 4 December 1901. Simon can barely keep his eyes open. He resists the urge to lay his head down on the desk in front of him, and instead tries to focus on Professor von Liszt, who's droning on about something at the front of the room.

It's a miracle he made it to class at all, Simon thinks, since last night turned into another marathon drinking bout with his friends. He staggered home at three in the morning, only to drag himself out of bed four hours later, stinking of stale ale and cigarettes. As he did so, he realized with dismay he was still drunk, a condition a strong cup of black coffee only slightly mitigated. Then he hurried out into the cold morning air.

Somehow he arrived on time, and now he's safely ensconced in the lecture hall, surrounded by the anonymity of twenty-four other faces. All he has to do is stay awake for an hour, and then he can crawl home to the warmth of his bed. If only the wooden chairs weren't so uncomfortable! Why can't the school turn the heat on? It's winter, after all!

Simon makes an effort to listen to the professor. He hears him mention the name of the French criminologist Gabriel Tarde. 'The fundamental forces that shape society, according to Tarde,' von Liszt is explaining, 'are imitation and innovation.' Simon decides listening is too much work. Instead, he focuses on the doodle taking shape in the margin of his book.

The professor pauses and acknowledges one of the older students, near the front of the room, whose arm is raised stiffly into the air. 'You would like to add something, Herr K.?'

Oh, no! Not that fool, Simon thinks. That man always has an opinion. He should shut up and let the professor finish so we can all go home!

'Yes.' Herr K. stands up to speak. 'I would like to examine Tarde's theories from the standpoint of Christian morality.'

Simon rolls his eyes as he continues to work on his doodle. He has no intention of listening to whatever Herr K. has to say, but his attention is brought back to the front of the classroom by a loud, sharp rap. The man sitting next to Herr K. has slammed his hand down on his desk. 'Not your Christian morality,' the man exclaims. 'I can't stand that!'

Simon smiles. Well put! Suddenly the lecture has become more interesting.

Herr K. spins around to confront his accuser. For a moment he stares at the man in outraged silence as a bright shade of apoplectic red creeps up his neck and spreads across his face. At last he spits out a response, 'Sir, you have insulted me!'

'Shut up!' his opponent shouts back. 'Every day you go on and on about your Christian morality. I'm sick of it. Don't say another word!'

Herr K.'s face flushes an even brighter shade of crimson. A blood vessel throbs violently on his temple. 'How dare you speak to me like that!'

What happens next occurs so rapidly that Simon can scarcely take it in. Herr K. rushes forward, knocking the desk in front of him to the ground. Suddenly there's a gun in his hand, pointed at the other man. He must have pulled it out from beneath his coat. His opponent leaps up and grapples with him. As the two men wrestle, the gun zigzags up and down, impossible to follow. Professor von Liszt rushes over to break up the fight. The three men struggle. It's all a tangle of arms and legs.

'Achtung!' somebody screams. 'Watch out!'

And then the gun goes off . . .

Given the culture of German student life in the early twentieth century, it's easy to imagine that a character such as Simon, the drunken student, might have attended Professor Franz von Lizst's winter-quarter criminology seminar at the University of Berlin. He is, however, a fictional creation. But the rest of the scene, including the heated verbal exchange, violent struggle, and gunshot, actually did happen.

However, there was a twist. After the gun went off, von Liszt stepped back, smoking gun in hand, and announced with a dramatic flourish that what everyone had just seen was a staged event – an experiment in applied psychology. The gun was loaded with blanks. The two men, Herr K. and his opponent, had faked their argument for the benefit of the class. The purpose of the experiment, von Liszt explained, was to explore the reliability of eyewitness evidence. Everyone in the classroom had seen the same confrontation, but would everyone's recollection of it be the same? To find out, von

Liszt asked the students to write down a description of what they had seen.

Still shaken by the conflict, the students obediently leaned over their desks and began writing. But as they searched their memories, it's doubtful any of them fully realized the significance of what they had just witnessed. They had doubtless all seen actors on a stage, but they would never before have seen an actor performing in a research study, because that was a scientific first. While experimenters had occasionally used deceptive methods in the past, none had ever crafted a setup that was deceptive in such an elaborate way.

To us here in the twenty-first century, conditioned by decades of TV shows like *Candid Camera*, it seems obvious that deception can be a useful (and often amusing) tool for studying human behaviour. But if you examine pre-twentieth-century science, it's rare to find examples of behavioural researchers purposefully misleading or concealing information from experimental subjects; and when you do find examples, the deception is typically quite minor. For instance, in 1895, Carl Seashore was conducting sensory research in which he asked subjects to hold a wire as he charged it with a weak electric current. 'Do you feel the current now?' he would ask. The subject would nod his head. On occasion Seashore asked this question even though he hadn't turned the current on, and he discovered that often his subjects nevertheless nodded to indicate they felt it. It was deceptive of Seashore to trick them into believing the current was on, but it was a little white lie. No one was going to get upset about it.

However, when people started fighting in von Liszt's classroom and a gun came out, that was deception of an entirely different magnitude. Carefully choreographed with actors and props, it more closely resembled a theatrical production than a scientific experiment. As the twentieth century progressed, the use of actors in psychological studies eventually became quite common, though researchers would refer to them as 'accomplices' or 'confederates' rather than actors, perhaps because those terms sound more formal and scientific.

What Was the Weather a Week Ago?

Although von Liszt conducted the experiment, he wasn't the one who came up with the idea. He merely provided the set. The roots of the experiment traced back eight years to March 1893, when Columbia University professor J. McKeen Cattell asked the fifty-six students in his lower-level psychology class to write down answers to four seemingly simple questions:

What was the weather a week ago today?
Do chestnut trees or oak trees lose their leaves earlier in the autumn?
Do horses in the field stand with head or tail to the wind?
In what direction do the seeds of an apple point?

The students had thirty seconds to write each response. This was in an era before junk food or automobiles, so Cattell assumed that horses and apples were things the students would have observed frequently in everyday life, but when he examined the answers, the results surprised him. Their responses were all over the map. For instance, their answers about the weather were 'equally distributed over all kinds of weather which are possible at the beginning of March'. Their guesses included 'clear', 'rain', 'snow', 'stormy', 'cloudy', 'partly stormy', and 'partly clear'. The correct information was that, a week ago, it had snowed in the morning and cleared in the late afternoon. The replies to all the other questions displayed similar confusion. And yet, when asked how sure they were about the accuracy of their responses, the students expressed strong confidence.

Cattell published his results in the journal *Science*, where they caught the attention of legal scholars and psychologists. It had long been known that eyewitness testimony wasn't very reliable, but Cattell's study suggested just how unreliable it might be. Highly educated Columbia University students, sitting in the relative calm of a classroom, struggled to remember last week's weather. So imagine how error-prone the memory of an eyewitness testifying in the high-pressure setting of a courtroom must be.

Cattell's study caught the attention of one young German psychologist in particular, Louis William Stern. Stern was a brilliant innovator, full of a restless energy that he applied to numerous different avenues of research. Early in his career he invented a *Tonvariator* or sound hammer – a device that allowed researchers to study people's sensitivity to subtle changes in pitch. Later he produced a study of Helen Keller, and invented the concept of the intelligence quotient, or IQ. That was all before he turned forty. But in the late 1890s, when he was still in his late twenties and read about Cattell's study, he decided to research the psychology of eyewitness testimony. He conducted a series of *Aussage* or 'remembrance' experiments. He showed subjects a picture and then asked them to describe details of what they had seen. Invariably, the subjects made numerous mistakes – even if they knew they were going to be asked about what they had seen, and even if Stern instructed them only to report those things they were absolutely certain about.

Stern's *Aussage* experiments convinced him that the legal system desperately needed reform. He was sure many innocent people had been sent to jail on the basis of faulty eyewitness testimony. But he quickly perceived that his experiments, as well as those of Cattell, were too abstract to make an impression on the legal community. Lawyers would simply dismiss them as artificial constructs of a research lab. What was needed, he decided, was something far more dramatic, something no one was going to forget, and which was undeniably true-to-life. He came up with the idea of staging a crime in a classroom. Stern reasoned that this would simulate, as closely as possible, the conditions under which a witness might view an actual crime. He called what he had dreamed up a *Wirklichkeitsversuch*, which translates loosely as a 'reality experiment'. Then he convinced his friend von Liszt to allow his classroom to be used as the setting.

Stern already knew about the unreliability of eyewitness testimony, so the results weren't a surprise. As expected, von Liszt's students produced accounts of the crime riddled with omissions, alterations, and outright errors. When describing who started the fight, the students offered eight different names. They disagreed about how long the confrontation had lasted, when exactly the gun had been fired, and how von Liszt had intervened. They garbled the conversation between the two men. Some even claimed that the man with the gun had fled the scene. But what von Liszt and Stern found most disturbing was that the most exciting events – the actual struggle with the gun – produced the greatest number of errors. In other words, the moment when the students were presumably focusing their attention most closely on what was happening was simultaneously the moment when their powers of recall became the most blurred.

Newspapers throughout the world reported on the unusual experiment. 'Pretended Quarrel As Classroom Test' ran the front-page headline in the *New York World*. The *Logansport Journal* opted for a more lurid description: 'Men Fight For Science's Sake'.

Both Stern and von Liszt used the publicity to argue for the necessity of serious legal reforms. 'What is to become of our entire criminal justice system if its surest foundation, the testimony of trustworthy witnesses, is shaken by rigorous scientific research?' von Liszt asked. The answer, he suggested, was to rely upon expert psychological advice in the courtroom. But this proposal didn't go over well with the legal community. Courtroom procedures followed centuries-old tradition. Judges and lawyers were pillars of the community. But what were psychologists? Mere upstarts. So why, lawyers asked, should they be granted a place of honour in the courtroom? Faced with this kind of resistance, the reforms went nowhere.

The Fighting Continues

Although the reality experiment didn't produce many changes in the courtroom, the same wasn't true of the classroom. Professors of law and psychology loved the experiment. Here, at last, was the solution to students falling asleep during lectures. Scare them awake! Suddenly there was no classroom or scholarly meeting safe from the sudden intrusion of a gun or knife-wielding stranger. During the years that followed, many twists were added to the experimental setup, but the basic plot always remained the same.

1906: During a meeting of a scientific association in Göttingen, a clown suddenly burst into the assembly hall, followed closely by a gun-wielding black man wearing a bright-red necktie and white pants. Both were shouting wild, incoherent phrases. The clown fell to the floor, and the black man leapt on top of him. A shot was fired, and then both men abruptly ran out of the room. The president of the association stepped forward. He alone knew the scene had been staged, but not wanting to tip his hand just yet, he asked the shocked attendees to write down what they had seen, in case the matter ever came before the courts. Forty reports were handed in. The esteemed scientists had all witnessed the intrusion at close hand, but they produced accounts that strongly disagreed on points as basic as what the men had been wearing, and how long the struggle had lasted. The president estimated that only six reports didn't contain statements that were positively wrong.

1914: At a meeting of a legal society in Vienna, the lawyer Franz Kobler stood up during a colleague's speech and began to berate him with insults. *You idiot! You moron!* Threats were exchanged. Kobler's companions thought he had gone mad. The offended colleague subsequently filed criminal charges against Kobler, and several tribunals interviewed eyewitnesses in order to render a verdict. Only then did Kobler reveal to everyone that the entire affair – the altercation and the tribunals – was a reality experiment. Analysis of the testimony showed that the eyewitnesses described the moments of calm before the dispute fairly accurately, but as the tension in the room mounted, their memories of events grew increasingly garbled.

1924: The psychologist William Marston was conducting a class at American University in Washington, DC, when a stranger knocked on the door and entered. The man walked over to Marston, handed him an envelope, and removed a long, green-handled knife from his pocket. He then turned to face the students and proceeded to use the knife to scrape menacingly at his gloved thumb. After that mysterious display, he left the room. Marston asked his students to write down what had just happened. Amazingly, and somewhat unbelievably, he claimed that not a single witness had noticed the knife.

1952: During a lecture at the London School of Economics, two members of the audience, an Englishman and a Welshman, began to quarrel. The Englishman stood up, drew out a gun, and shot the Welshman. The audience sat stunned until Professor Laurence Gower revealed it had all been staged and asked everyone to describe what they had seen. Once again, the reported details of the scene differed significantly.

By 1975, the psychologists Robert Buckhout, Daryl Figueroa and Ethan Hoff, writing in the *Bulletin of the Psychonomic Society*, observed with a hint of weariness that 'Would-be "criminals" have been running into psychology classes for decades, committing "crimes" and creating eyewitnesses, eyewitnesses who later prove to be unreliable and inaccurate.' Nevertheless, they continued, 'As compelling as this demonstration of poor recall is, eyewitness testimony continues to be overrated in the courtroom and is the source of many convictions of innocent people.' So they staged yet another reality experiment, arranging for a student at California State University, Hayward, to 'attack' a professor in front of 141 bystanders. Again, eyewitness testimony proved to be error-prone.

During all these decades of simulated classroom crimes, it never seemed to occur to anyone that the experiment was potentially dangerous. What if someone had a gun and used it against the pseudo attacker? What if the fake fight started a real fight? What about the feelings of the students forced to witness the confrontation? Today it's impossible not to ask those questions, because classroom violence – genuine, not simulated – has become a distressingly common occurrence, making the reality experiment seem like an odd relic of a more innocent age. In fact, it seems unthinkable that in the twenty-first century a teacher would purposefully stage a violent confrontation as a classroom exercise.

And yet scenes reminiscent of the reality experiment continue to pop up in the news with surprising regularity. In 2004, armed intruders shouting, 'There will be no new taxes! Everyone here is going to vote no!' burst into a meeting of government officials in Carter County, Tennessee. It was an unannounced terrorism 'drill' cooked up by the local Emergency Management Director. In 2007, young students at Scales Elementary School in Tennessee, having been warned about a gunman loose in the area, dove for cover when a man wearing a hooded sweatshirt tried to break into their classroom. Their teachers later revealed they had thought it would be a good idea for their pupils to learn what a hostage situation might feel like. And in March 2010, a man appeared out of nowhere and 'shot' a science teacher in front of horrified students at Blackminster Middle School in Worcester. It turned out he was participating in a role-playing exercise designed by the faculty to teach the kids how to 'investigate, collect facts and analyse evidence'.

For students, the implication of these continuing walk-on appearances by phoney assailants is clear: it's still not safe to sleep in the classroom.

The Psychologist Who Hid Beneath Beds

Bryn Mawr College, Pennsylvania – 1938: 'I don't think we should be here.'

'Shhh! Be quiet.'

'But what if they see us?'

'They won't! And remember, we're doing scientific research.'

'But . . .' The woman on the left clamps a hand over her nervous companion's mouth. Just at that moment a door swings open, and the two women hurriedly inch a little further beneath the bed they're hiding

under. They hear footsteps approaching. Several pairs of women's legs come into view. Then there's the sound of voices.

'Betty, I'm really looking forward to this party.'

'So am I. I just hope we have enough food for everyone.'

'I think we should. Let's see. Cookies, potato chips, pie . . .'

Beneath the bed, the two researchers begin to take notes, but since they're lying on their stomachs, trying not to make any noise, it's difficult to find a comfortable position in which to write. They awkwardly shift their weight from one elbow to the other. Above them they hear the sound of objects being arranged on a table. Then there's more conversation.

'Did you hear who Sarah is dating?'

'No, tell me!'

'Brad!'

'Oh my God! I don't believe it.'

'He's so cute.'

One of the researchers flips a page in her notebook. As she does so, she fumbles and drops her pencil. It makes a sharp rap as it hits the hardwood floor. Immediately, the voices stop. The researchers freeze in place, holding their breath, praying they won't be discovered. A few seconds that feel like minutes pass in silence. The women's hearts pound in their chests. Finally, above them:

'Betty, did you hear something?'

'I did. Is there someone else here?'

A long pause. 'I think it may be mice. You know, there's lots of them in these old buildings.'

'Ewww. How disgusting. I'll tell the janitor to put out some traps.'

The beginnings of covert observation, as a technique of scientific inquiry, were innocent enough. Every evening at around 7.30 p.m., during the spring of 1922, Professor Henry T. Moore would leave his house in New York City and take a long, slow stroll along the stretch of Broadway known as the Great White Way, because of the brilliant marquee and billboard lights that illuminated it. He walked behind couples arguing, stood beside businessmen chatting at bus stops, and waited outside theatres to hear people coming out from the shows. And always, he jotted down the remarks he heard in his notebook, though often he had to strain to hear what people were saying over the noise of traffic rumbling past.

Moore's reason for this systematic eavesdropping was to study whether men and women emphasized different subjects in their everyday conversations. Among other things, he discovered that women talked about men far more often than men talked about women. This use of surreptitious surveillance was deceptive, but only mildly so. After all, his subjects were in a public location. They had to assume other people might overhear what they were saying. Moore merely took advantage of this assumption for the sake of his scientific research.

Two years later, Carney Landis and Harold Burtt extended Moore's research by eavesdropping at a wider variety of locations. Wearing rubber-heeled shoes and cultivating an 'unobtrusive manner', they loitered around Columbus, Ohio railroad stations, department stores, and hotel lobbies, taking notes on every conversation they overheard. Like Moore, they concluded, 'Persons play a small part in man's thought and a large part in woman's.'

It was in 1938 at Bryn Mawr, a women's liberal arts college in Pennsylvania, that the technique of covert listening advanced to its next logical stage. Mary Henle was conducting research there for her doctorate in psychology. She had been intrigued by a hypothesis once made by the child psychologist Jean Piaget, who had observed that children, when talking, make a large number of references to themselves. Attributing this self-absorption to childhood 'egocentricity', Piaget theorized that as children grew older and became more socialized, they would cease to be so inwardly focused, causing the number of self-referential remarks made by them to decline. Henle decided to put this theory to the test by surreptitiously listening in on the conversations of adults and recording the number of times they referred to themselves. Luckily she had a large pool of adults ready at hand to eavesdrop on – her fellow Bryn Mawr students.

To conduct her research, Henle enlisted the help of her friend Marian Hubbell, and together they launched an all-out spy operation. But unlike previous researchers, they didn't limit themselves to collecting data in public areas. Instead, they extended their investigation into the most private of locations. As they put it, 'Unwitting subjects were pursued in the streets, in department stores, and in the home.' They crouched down in the bathroom stalls of the women's dormitory to overhear washroom gossip; they lifted up phone receivers to monitor intimate discussions; and they snuck into the rooms of their fellow students and hid beneath their beds.

Spying on young women in dorm rooms is the stuff of prurient fantasy, so one can't help but wonder what the two researchers might have seen or overheard. Were they privy to whispered confessions of sexual secrets shared by room-mates? Did they overhear anything potentially incriminating? Also, how long did Henle and Hubbell hide beneath the beds? Were they ever caught? Unfortunately, our curiosity about these matters must remain unfulfilled, because the researchers didn't share many details. They merely referred to the data-collection process as 'difficult', and left it at that.

As for their results, it turned out Piaget was wrong. Adults, at least those eavesdropped on by Henle and Hubbell, talked about themselves just as much as children did. The number of self-referencing remarks for both groups came in at around 40 per cent. But really, who cares? (Except for a few psychologists.) Henle and Hubbell's study is definitely one of those in which the method of inquiry was far more interesting than what was being studied. In fact, the study would probably have sunk quietly out of sight if not for the unorthodox research technique. Instead, it's earned a minor but recurring place in textbook discussions of research ethics, accompanied by warnings that hiding beneath beds is not a recommended method for conducting fieldwork.

Midnight Climax

Henle and Hubbell may have been too discreet to report on all the juicy details of their research, but it was only a matter of time before someone whose motives were less pure used covert observation in a more salacious way. That moment arrived in 1955, when a bordello opened in San Francisco. On the surface it seemed like any other of the city's similar establishments. The interior was garishly decorated. Pictures of cancan dancers and dominatrixes hung on the walls. Sex toys could be found in every drawer, and the drinks flowed freely. But what the johns who accompanied the prostitutes inside couldn't have known was that every move they made in there was being observed by CIA-employed psychologists hiding behind one-way mirrors. Nor did they know their drinks had been spiked with LSD, though they probably guessed something had been slipped to them when they woke up the next morning after what must have seemed like the strangest night of their lives.

The CIA called it Operation Midnight Climax. Its purpose was to provide the agency with real-world psychological data about the use of LSD on unwitting subjects – such as whether the drug could be used as a truth serum or as a brainwashing tool – as well as to give their agents a chance to hone their sexual blackmail techniques. American taxpayer dollars at work!

Just as with Henle and Hubbell's research, it would be interesting to know exactly what the CIA researchers saw and overheard – the dizzying spiral of their subjects' minds into LSD-laced psychosis – but those details remain a state secret. The operation shut down in 1963, and by the time its existence was publicly revealed during Senate hearings in the mid-1970s, all the relevant documents had been heavily censored. It's not even known what became of the unwitting participants, or what long-term effects they might have suffered.

Lovers, Friends, Slaves

Of course, the CIA is in the business of spying, which doesn't make its actions in Midnight Climax acceptable, but it does make them not entirely surprising. That same excuse, however, doesn't apply to the social worker Martha Stein.

Concerned that there was a lack of scientific information about the behaviour of the male customers of call girls, in 1968 Stein undertook a four-year study of this subject. With the cooperation of sixty-four New York City call girls, who apparently 'found it gratifying that a well-educated researcher respected them and considered their work important', Stein started spying on the full range of activities that occurred between prostitutes and their johns. In many cases, the call girls already had one-way mirrors and peepholes installed in their apartments, either for the benefit of voyeuristic clients or to allow them to observe a girl-in-training. In such cases, Stein's job was easy. But at other times, Stein had to hide in closets, furtively peering out through a crack in the door to see what was happening. The call girls assisted by making sure their clients were facing away from her.

Unlike Henle and Hubbell and the CIA researchers, Stein wasn't shy about publicly revealing every explicit detail she observed. Thanks to her careful research, we know that of the 1,230 men she observed, 4 per cent were cross dressers, 11 per cent asked for threesomes, 17 per cent wanted to be tied up during sex, 30 per cent enjoyed anal stimulation, 36 per cent French-kissed the call girls, and almost all of them wanted fellatio. Her publisher, hopeful that such frankness would translate into brisk sales, heavily promoted her book (*Lovers, Friends, Slaves: The Nine Male Sexual Types*) in newspapers such as the *Washington Post, Chicago Tribune*, and the *New York Times*. Readers were promised a voyeuristic peek through 'see-through mirrors to watch more than 1,200 men in their sexual transactions'. Stein fully delivered.

Of course, this brief history of the use of covert observation in scientific studies shouldn't scare us into paranoically checking beneath our beds or in our closets. Most psychologists, we can be sure, aren't constantly spying on their neighbours. However, if you do find yourself in an unfamiliar setting such as a hotel or dorm room, a quick check around for any errant researchers might not be a bad idea.

The Metallic Metals Act

New York City, New York – 1947. The phone rings as Susan is preparing dinner.

'Honey, can you get that?' she calls out.

'I'm busy!' a male voice replies from the living room.

The phone rings again, its urgent cry demanding a response. Susan glances over at young Benjamin sitting in his high chair. He's carefully examining his food and seems content for the minute. She turns the stew bubbling on the stove down to a low simmer, steps over to the phone, and lifts it from its cradle just as it rings a third time.

'Hello?'

'Hello, ma'am. I'm conducting a public opinion survey on behalf of the Sherman & Marquette agency. I was wondering if I could have a minute of your time.'

Susan looks at the stove. 'I'm a little busy now. I was just preparing dinner.'

'This will be very brief. I promise.'

She sighs. 'Well, I suppose.'

'Thank you, ma'am. This survey has only one question. We'd like to know which of the following statements most closely coincides with your opinion of the Metallic Metals Act? a) It would be a good move on the part of the US; b) It would be a good thing but should be left to the individual states; c) It is all right for foreign countries but should not be required here; or d) It is of no value at all.'

Susan pauses. 'Um. Could you repeat the question?'

'Certainly, ma'am.' The caller repeats the question as well as the four options.

Susan thinks for a moment. 'Well, I guess I would say B.'

'Option B? That the Metallic Metals Act would be a good thing but should be left to the individual states?'

'Yes, that's my answer.'

'Thank you, ma'am. I'm sorry to have disturbed you during dinner.'

'Oh, it's not a problem. Goodbye.'

'Goodbye.'

Susan has a slightly puzzled look on her face as she carefully places the phone back on its cradle, but then she shrugs and returns to stirring the stew. A minute later her husband strolls into the room holding a folded newspaper beneath his arm. 'Who was that on the phone?' he asks.

'Oh, no one,' Susan replies. 'Just some public-opinion survey.'

In March 1947, Sam Gill, research director of the Sherman & Marquette ad agency, reported a curious finding in *Tide*, a trade journal of the advertising industry. He asked subjects the question posed by the caller in the scenario above: what was their opinion of the Metallic Metals Act? He reported that 70 per cent of those surveyed readily offered an opinion. Of this group, '58.6% favored leaving the Metallic Metals Act to individual states; 21.4% thought it would be a good U.S. move; 15.7% thought it shouldn't be required and 4.3% thought it had no value at all.'

The curious part of this finding was that the Metallic Metals Act didn't exist. It was an entirely fictitious piece of legislation. But many people apparently had an opinion about it nevertheless. Gill suggested the response to his survey demonstrated that 'the average U.S. "man on the street", while not the world's worst liar, is always willing to give an "opinion" on any subject whether he knows anything about it or not'.

Uninformed Opinions

Gill's study has become something of a classic, frequently cited in discussions of survey methodology as a reminder that the responses of subjects are not always well informed or meaningful. If you ask a person a question that appears, on the surface, to be sensible, often he will oblige you with a reply, even if he has no idea what he's talking about. There are a number of possible reasons why this happens. Perhaps the person feels pressured to provide some kind of answer. Perhaps he's embarrassed and doesn't want to admit ignorance. Or perhaps he confused the question with another issue that sounds similar, about which he has a legitimate opinion. Subsequent research has indicated that if surveyors provide a 'no opinion' or 'don't know' option it lessens the number of nonsense replies, but it doesn't eliminate them altogether. Some people, it seems, just like to offer opinions – any opinion at all!

Gill's experiment offers a well-known example of the uninformed opinion phenomenon, but it's not the only one. The use of fictitious questions to trigger nonsense responses is a minor but persistent genre within public opinion and sociological research.

One of the earliest examples of the technique comes from 1946, a year before Gill conducted his study. Eugene Hartley was surveying American college students about their attitudes towards various foreign nationalities. He asked how they felt about the French, Italians, Mexicans, Chinese, etc. Should members of these groups be allowed into the country? Would you want one of them as your neighbour? Would you marry one of them? While doing this, he had the idea of including three 'nonesuch' groups in his list: the Wallonians, Danireans, and Pireneans. He discovered that many of the students quite readily expressed an opinion about these non-existent nationalities. In particular, if a student already was inclined to be intolerant of foreigners, he definitely didn't want any of those Wallonians, Danireans, or Pireneans entering the country.

It's more common for researchers to elicit opinions about fictitious pieces of legislation and government organizations. For this reason, we know that the public has definite views on bogus issues such as the Religious Verification Act, the 1975 Public Affairs Act, and the National Bureau of Consumer Complaints.

In 1976 several Oxford University researchers discovered that people would also readily provide feedback about invented geographic locations. While travelling through Iran they systematically asked strangers for directions to the non-existent Hotel America in Tehran and to the Hotel Abadan in Isfahan. The strangers – at least, those who were willing to talk to them – quite happily provided them with detailed directions that would have led any real tourist on a wild goose chase. Just to make sure this wasn't a case of mischief towards foreigners, the researchers simultaneously asked for directions to a real and well-known place, to which request accurate replies were given.

The Oxford researchers later repeated the experiment in England by posing as foreigners and asking for directions to the spurious Hotel Hazel Grove. The English, it turned out, were far less likely to provide phoney directions, which led the researchers to hypothesize that there was something about Near Eastern culture that made the people there peculiarly eager to *appear* knowledgeable and helpful, even if this appearance had no basis in reality: 'If it is assumed that the act of giving directions to a fictional place reflects a greater concern with form rather than substance, then the data gathered in this study may be taken to show that significantly more Iranians than English people value form over substance.'

There's a Pattern Here

The uninformed opinion phenomenon has implications far beyond the design of surveys or giving directions to tourists. Voters, for instance, have proven themselves willing to cast votes even when they don't recognize any of the candidates running for office. Consumers also find themselves frequently forced to choose among products and service providers, none of which they know anything about. For instance, have you ever hired a plumber by randomly picking a name from the phone book? Or have you stood in the grocery store deliberating between brands, all of which are equally unfamiliar?

The sociologist Stanley Payne realized that people don't respond in a purely random fashion when confronted with a range of unfamiliar options. There are patterns to their uninformed choices. First, they exhibit a tendency to seek any kind of 'middle ground' – which is probably why a majority of Gill's respondents gravitated towards the ambivalent option of leaving the issue to the individual states. Second, they read meaning into unknown phrases based upon their similarity to known phrases. Third, they often opt for the last choice given. Finally, and above all else, they seek out the 'appeal of the familiar'. If a person recognizes anything at all about one of the options, that's the one she'll probably choose. Advertisers are very aware of this, which is exactly why they spend so much money parading their products before our eyes. They're hopeful that when we walk down a grocery aisle our hands will instinctively reach for the product we vaguely remember having seen on TV.

There's a final twist to the story of the Metallic Metals Act. In 1978, the researchers Howard Schuman and Stanley Presser decided to dig up Gill's original report in *Tide* magazine. After all, the experiment was widely cited, but no one appeared to have seen Gill's actual data. Eventually they located the appropriate issue of *Tide* – not an easy task, since the number of libraries that archive the magazine can be counted on one hand. What they found was a disappointment. Gill provided almost no details about the study beyond the numbers already given. He didn't say how he conducted the study (did he phone people or interview them in person?), nor when he did it, nor how many people he questioned. Schuman and Presser concluded the experiment was 'hardly more than an anecdote'. In fact, there's no evidence beyond Gill's word that he even conducted the experiment. His famous account of uninformed responses to a fictitious question could itself be a fiction, which adds an ironic twist to his study.

However, even if Gill simply made up his results (which we don't know for sure), his basic insight nevertheless appears sound – especially since other researchers have been able to demonstrate it, specifically with regards to the Metallic Metals Act. In 1981, the marketing researchers Del Hawkins and Kenneth Coney mailed a questionnaire to 500 people randomly chosen from telephone directories in Portland, Phoenix, Cincinnati, and Buffalo. In a nod to Gill, their survey included the following question: 'Passage of the pending Metallic Metals Act will greatly strengthen the economic position of the United States. Yes or No?' They reported that the majority of their respondents felt that the Act would strengthen America's economy. So if in the future you hear American politicians hammering on about the regulation of metallic metals, you'll know why. Because if it's true that the Man on the Street will happily express an opinion on a subject whether he knows anything about it or not, it's even truer that politicians seeking office will eagerly pay lip service to any issue, meaningful or not, that polls well.

A Roomful of Stooges

Swarthmore College, Pennsylvania – 1951. Jason sprints down the corridor towards the classroom. He stops directly in front of it, pauses to catch his breath, and then knocks on the open door. 'Is this the vision study?'

Six young men are seated inside around a square table. A thin, balding man wearing a grey jacket and tie stands at the front of the room.

'You've come to the right place,' the man replies with a smile. He has a gentle voice, with a hint of an Eastern European accent. 'Please come in.' He gestures towards an empty chair.

'Sorry I'm late,' Jason says as he takes his seat. 'The bus was really slow.'

'Not a problem. Thank you for coming. I was just introducing myself to the other volunteers. I am Professor Asch. I work in the psychology department here at Swarthmore. And you are all here because you responded to advertisements I placed in newspapers at your colleges.'

Asch pauses to pick up two large cards from a stack on a desk beside him. He places the cards on a stand so they're visible to the volunteers. One of the cards shows three vertical lines of different lengths, numbered 1, 2, and 3. The other card shows a single vertical line.

Asch continues, 'As the advertisement stated, this is a psychological experiment in visual judgement. The procedure is quite simple. I don't expect it will take more than half an hour of your time. I am going to show you a series of paired cards. The card on the left,' he points at the card, 'will always show a single vertical line. The card on the right will always show three vertical lines of different lengths. I want you to choose from the card on the right the line that is the same length as the line on the left-hand card.'

He pauses to let this sink in. Jason looks at the cards. He can see immediately that line 2 is the same length as the single line.

'You will each state your choice in the order in which you are seated at the table. Do you understand?' Everyone nods. 'Good, then let's begin.' Asch gestures at the first volunteer immediately to his left, a clean-cut man in a neatly pressed shirt. 'Please start.'

'Line 2,' the volunteer says. The next volunteer echoes this choice, as does the next, and the next. Jason is seated second from last. When his turn arrives, he also says, 'Line 2.' This is going to be easy, he thinks.

The next round proceeds in a similar manner. All seven volunteers choose line 1. What a pointless experiment, Jason thinks.

Then round three begins. Asch places new cards on the stand and nods at the first volunteer.

'Line 1,' the young man says. Jason blinks. Line 1? He looks more carefully at the cards. Line 3 appears to be the correct answer. In fact, it quite obviously is. Assuming the volunteer made a mistake, Jason waits for the next person to correct him. Instead, the second volunteer also says, 'Line 1,' as does the next, and the next.

Jason feels confused. Did he mishear the instructions? Why is everyone giving the wrong answer? 'Line 1,' the student next to him says, and then it's his turn. He has no time to consider the situation. Everyone else can't be wrong, he thinks. I must have misunderstood the researcher. 'Line 1,' he blurts out.

The person beside him shifts in his chair. For a second, Asch glances down. Jason's stomach tightens slightly. He only dimly hears the volunteer to his left also saying, 'Line 1.' Then they're on to round four.

The same mysterious phenomenon occurs again. As Asch places the new cards on the stand, Jason sees that the answer is line 2, but the first volunteer says, 'Line 3,' as do the others. When his turn arrives, Jason thinks: I don't know what's going on, but I don't want to ruin this guy's experiment. He picks line 3.

Three more rounds pass. Each time, as before, the other volunteers pick a line that seems obviously to be the wrong answer, and each time Jason echoes their choice, though his sense of unease deepens every time he does so. He wonders how he could have misunderstood the instructions so badly. They sounded so easy when he first heard them. Were they supposed to compare line width? Or perhaps there's some sort of optical illusion involved? But neither of these explanations makes much sense. It wouldn't be so bad, Jason considers, if the other volunteers didn't sound so confident when they gave their answers.

It's now round eight. Asch places new cards on the stand. The answer, Jason sees, is line 2, but it doesn't surprise him when the first volunteer says, 'Line 1.' Nor is he shocked when volunteers two through six also make this choice. His turn arrives again. An uncomfortable, sinking sensation has crept over him. He wishes the experiment were over. He also wishes he understood what was going on.

Everyone is waiting for him to speak. He can feel their eyes focused on him, sizing him up. He looks at the cards. Line 2 is the obvious answer.

'Line 1,' he says.

By the 1950s, the residents of the United States had become acutely aware of the dangers of 'conformity'. The word evoked images of totalitarian control – Nazi propaganda and Chinese brainwashing. But as Americans looked around at their own country, they worried that disturbingly similar signs of groupthink were creeping into their society – that the rugged individualism that formed such a basic part of the American self-image was slipping away. Wasn't it a little unsettling how young people all strove to look and talk alike, each sporting similar sweaters, rings, and school pins? Didn't businessmen commuting to work resemble, just slightly, an army of clones dressed in a uniform of grey flannel suits? In interviews conducted by the *New York Times* in 1958, American college students overwhelmingly said that they anticipated their most deep-felt personal problem in life would be the conflict between their desire for success and resisting 'society's pressures for conformity'.

An Experiment in Conformity

Solomon Asch was particularly aware of the pressures and dangers of conformity. He was born in a small Jewish community in Poland but moved to America in 1920, at the age of thirteen. As a teenager living on New York's Lower East Side, he struggled to assimilate into American society, but he succeeded, and by the 1940s he was a professor of psychology at Brooklyn College. Then he watched as war ravaged his former homeland and his people. His response was to study the Nazi techniques of propaganda and indoctrination, hoping to understand how the Nazis had succeeded in coming to power and manipulating the German public. Unlike many of his colleagues, however, he refused to believe that individuals were helpless in the face of group pressure – even when that pressure was organized by the state. He felt sure individuals had the strength within them to resist, and in 1951, after moving to Swarthmore College in Pennsylvania, he decided to put his conviction to the test.

Asch's Jewish heritage served as his inspiration. One day he was thinking back to a Passover dinner in Poland, when he was seven. As was the custom, an extra cup of wine had been placed at the table for the prophet Elijah. 'Just watch,' an uncle had leaned over to tell him. 'Elijah will take a sip of it.' Young Solomon eagerly watched the glass throughout the entire meal, hoping to see the moment when the invisible spirit of Elijah arrived and drank the wine. Finally he convinced himself that the level of the liquid had dropped slightly. Elijah had visited!

Thinking back on that experience as an adult, Asch realized that the power of suggestion had made him see something that never occurred. Would group pressure, he wondered, similarly be able to alter what an individual saw – or at least what he claimed to see? For instance, if a group of people claimed two lines were of equal length, even if they obviously weren't, would their unanimous opinion be persuasive enough to compel an unwitting subject to agree? What would prove more compelling: the evidence of their own senses, or the pressure to conform to the group? Asch decided to find out.

Like a spider luring flies into its web, Asch placed an ad in the student newspaper of nearby Haverford College, promising volunteers a small financial reward if they participated in a 'psychological experiment in visual judgement'. One by one, victims wandered into his trap.

The experiment proceeded much as described in the introductory scene. A Haverford student showed up at the appointed time and found a group of other volunteers already waiting in a room. Thinking nothing of it, he took his seat. Then Asch walked in and explained that the task of the volunteers was to compare line lengths. It must have sounded easy, almost ridiculously so, but quickly the experience turned surreal for the Haverford volunteer as all the other subjects jointly began to give incorrect answers – which wouldn't have been so bad, except that they all gave the same incorrect answer. The psychologist Roger Brown later described the setup as an 'epistemological nightmare' because it offered subjects a stark choice: am I going crazy, or is everyone else?

Of course, what the confused Haverford student didn't know was that he was the only true subject of the experiment in the room. The other young men seated around the table were stooges – Swarthmore students in collusion with Asch. Asch had coached them to give incorrect responses in twelve out of eighteen rounds of the experiment. So if the Haverford student felt a paranoid suspicion that all eyes in the room were focused on him, that's because they were.

Asch's research associate Henry Gleitman, who occasionally helped conduct the experiment, many years later described how emotionally wrenching the testing could be. When volunteers started echoing answers that were obviously wrong, it was like watching a slow-motion train wreck. It was hard not to cringe. 'You are ashamed for him,' Gleitman said, 'and I have the sense of embarrassment I have when I see an actor who blows his lines. I want to sink through the floor with him.' Gleitman recalled that some of the conforming subjects, when finally told the truth, broke down in tears – perhaps tears of relief that they weren't crazy.

Asch had assumed most people would defy the group pressure. But the results belied such hopes. Asch tested 123 individuals. Of this number, a full 75 per cent succumbed to group pressure at least some of the time; 25 per cent conformed over 50 per cent of the time; and a hard-core group of conformists, about 5 per cent of the total, always agreed with the majority, no matter how obviously incorrect its opinion was.

But even the non-conformists, those who consistently resisted group pressure, didn't do so confidently. They hemmed and hawed. They leaned forward in their chairs and squinted at the cards. They apologized profusely. 'I'm sorry, guys,' they would say, 'I always disagree.' And when asked if they thought the others were wrong (before the truth had been revealed), they were reluctant to say this. Instead, they preferred the less confrontational phrase that they *saw things differently*.

Asch conducted numerous variations on the experiment. He found that the presence of a supporting partner – someone else willing to contradict the group – greatly increased non-conformity. Also, the conformity effect only truly kicked in when individuals faced groups of three or more people. But overall, the results rattled Asch. He gloomily noted: 'That we have found the tendency to conformity in our society so strong that reasonably intelligent and well-meaning young people are willing to call white black is a matter of concern. It raises questions about our ways of education and the values that guide our conduct.'

The Golden Age of Deception

Asch published an account of his study in *Scientific American* in November 1955. It immediately fuelled new concerns, especially among educators, about America's drift towards conformity. Brown University President Barnaby Keeney told his students several weeks later, during a weekly chapel service, that he wanted to see less conformity among them. He jokingly suggested they should express their individuality by doing something radically different, such as tidying their dorm room. At Harvard, the theologian Paul Tillich managed to find a biblical admonition of conformity that he shared with his congregation: 'And be not conformed to this world, but be ye transformed by the renewing of your mind'. (Romans 12:2.)

But it was among psychologists that Asch's study had the most profound effect. By 1955, there had been numerous studies that used deception, but no one had ever used it on such a grand scale. No one had ever packed an entire room full of actors as Asch did. His colleagues were deeply impressed and rushed to design ingenious deceptions of their own. As a consequence, the next two decades became the Golden Age of Deception in psychological research. Deception became sexy and prestigious. If you didn't do it, you weren't part of the psychological 'in crowd'. By the early 1970s, over half of all the articles published in psychology journals reported the use of deception, having risen from only 20 per cent in the early 1950s. Of course, there was a touch of irony in the fact that a study on conformity helped inspire this fad for duplicity.

Numerous studies, including some famous in their own right, were directly inspired by Asch's experiment. For instance, Stanley Milgram, who had once worked as Asch's teaching assistant and had conducted a modified form of the conformity study for his doctoral dissertation, wondered what would happen if something more consequential than comparing line lengths was at stake. What if a person was pressured to do something morally repugnant, such as delivering a fatal electric shock to an innocent victim? Would he or she then be as willing to comply?

To explore this question, Milgram, like Asch before him, recruited volunteers by running an ad in a newspaper offering a small payment in return for participation in an innocuous-sounding 'study of memory and learning'. But he departed from Asch's study by doing away with the roundtable panel of stooges and using in its place the authority of a researcher – really an actor in a white lab coat – who directed a subject to deliver increasingly strong electric shocks to a victim. The researcher was supposedly studying whether the threat of shocks would aid memorization.

The shocks were fake, but the unwitting volunteer, listening to the agonized cries of the victim, didn't know that. Whenever the volunteer expressed hesitation, the pseudo researcher cryptically stated, 'The experiment requires that you continue.' To Milgram's surprise, two-thirds of his subjects unquestioningly accepted this command and kept pressing the button to shock the victim, even past the point at which the victim appeared to die. Milgram's 'obedience study' is arguably the most famous psychological experiment of the twentieth century.

At Columbia University, psychologists John Darley and Bibb Latané wondered whether the instinct for self-preservation would override the conformity effect. That is, would people conform to the behaviour of a group if doing so appeared to place them in danger?

Darley and Latané's volunteers believed they were going to participate in a discussion on the problems of urban life, but the researchers told them that a few forms first needed to be filled out and then led them to a room where several other people were already seated, busily completing questionnaires. The volunteers got to work, but after a few minutes smoke began to enter the room through a small vent in the wall. By the end of four minutes, there was enough smoke to obscure vision and interfere with breathing.

The researchers had constructed a system to pipe smoke into the room, but the unwitting volunteers didn't know that. As far as they knew, the smoke was evidence of a fire. Invariably, the first thing they did was to look around the room to check everyone else's reaction, but since the other people in the room were the researchers' secret accomplices, they weren't reacting at all. The other people merely looked up at the smoke and shrugged their shoulders. If the panicked volunteers asked them about it, they said, 'I dunno,' and continued working. It was up to the lone volunteer to take action and report the fire. Darley and Latané described what happened next:

Only one of the ten subjects . . . reported the smoke. The other nine subjects stayed in the waiting room for the full six minutes while it continued to fill up with smoke, doggedly working on their questionnaires and waving the fumes away from their faces. They coughed, rubbed their eyes, and opened the window – but they did not report the smoke.

Conformity had easily bested self-preservation.

The influence of Asch's study also extended to some more obscure studies. The researchers at the McCormick Corporation, makers of spices, herbs, and other flavours, had for years been using preference panels to test the palatability of products. But after reading about the conformity experiment, it occurred to them that one or two people with unusual tastes could potentially sway the opinion of the entire panel. Hurriedly, they decided to get to the bottom of the situation. On a five-person mayonnaise preference panel, they surreptitiously placed several 'mayonnaise stooges' who had been coached to express partiality for flavours such as 'meaty', 'lemon', or 'metallic mustard'. To their dismay, the food researchers discovered that 'extreme observations by an individual panel member' did indeed sway the judgement of the other panellists. Steps were promptly taken to address the problem.

However, most Americans weren't exposed to the conformity effect through any psychology study, even Asch's. Instead, it was through one of television's most popular shows, *Candid Camera*. Allen Funt, the creator of the show, had some training in psychology. As an undergraduate at Cornell University he had worked as a research assistant in the psychology department. The skits often reflected this background.

A 1962 segment titled 'Rear Facing' opened by showing a bald-headed man wearing a black trench coat standing in an elevator. Funt's voiceover identified him as the 'candid star', the unwitting victim. Other passengers, a woman and two men, entered the elevator – all of them *Candid Camera* actors. But instead of facing the door, as is the custom, they stood facing the back wall. The man in the trench coat looked around, confused. Funt's voiceover supplied the psychological narrative: 'You'll see how this man in the trench coat tries to maintain his individuality.' The man rubbed his nose. He looked at his watch. He glanced back and forth at the other passengers. And then, slowly, unable to resist the silent pressure of the group, he turned to face the wall.

Subsequent scenes repeated the gag with other victims. A businessman, with a puzzled look on his face, turned to face the wall almost immediately. Next Funt's actors succeeded in making a wide-eyed young man turn in a complete circle. 'Now we'll see if we can use group pressure for some good,' the voiceover said. All the actors removed their hats. The young man promptly did so as well.

The skit is considered one of the show's all-time classics. And, of course, its illustration of the almost irresistible power of group pressure is a page taken directly from Asch.

Versions of Asch's conformity experiment have been repeated hundreds of times since the 1950s, in almost every country in the world including Kenya, Fiji, Zimbabwe, Kuwait, New Guinea, and among the Eskimo of Baffin Island. Some variations have been found between cultures. The British, for instance, turn out to be rather non-conformist, as do, more surprisingly, the Japanese. But overall the conformity effect appears to be quite robust across all of human society. However, there is some intriguing evidence, detected by a meta-analysis of 133 studies spanning 50 years, suggesting that the effect has been weakening over time. People nowadays seem to be more willing to defy the authority of groups. Whether this is a good thing, or the beginning of a slide into anarchy, remains to be seen.

Thud!

A mental hospital in rural Pennsylvania – 1972. Robert's bedroom is a narrow cell with a single small window and an old army-style cot. The smell of hospital disinfectant hangs thick in the air. He sits on the edge of the cot, staring at the peeling paint and discoloured floor tile. Then he picks up his notebook and begins to write:

DAY ONE

Thought it would be harder to fake my way into a mental hospital, but here I am. Butterflies in my stomach when I arrived at the admissions office. Was sure I'd be exposed. Shouldn't have worried. The conversation went something like this –

Attending Physician: ' What's troubling you?'

Me: ' I'm hearing voices.'

' Voices? What kind of voices?'

I looked directly in his eyes to appear as sincere as possible.' A male voice.' A pause for dramatic effect. ' It says Thud and sometimes Hollow.'

He held my gaze for a moment and neither of us moved. I was afraid he wasn't going to buy it, but then he picked up his pen and jotted a note on a pad of paper. I guess he bought it.

' Thud and Hollow,' he repeated.

' Yes. And on occasion Empty.'

He nodded, as if this made sense.

' When did the voices start?'

' About three weeks ago.'

' And you find the voices disturbing?'

' Well, naturally, I'm concerned.'

' Have you talked to a physician about this?'

I shook my head.' I don't know any doctors in the area, but my friends told me this is a good hospital, so I decided to come directly here.'

He nodded again. The conversation shifted. We talked for a while about my family life. I told him the truth. I'm closer to Mom than Dad, but the relationship with Dad is improving. He found this interesting. Next thing I knew, I was admitted. It was that easy!

First impression of the facility: clean but shabby. Decaying around the edges.

No idea what to expect. How long will I be in here? Days, weeks, months? Suddenly nervous. What have I gotten myself into?

A heavy door bangs shut somewhere close by. Robert looks up from his writing. Footsteps approach from the end of the hallway. They stop near his door. Then the voice of an attendant rings out, harsh and mocking. 'Lights out, muthafuckas!'

Abruptly, all goes dark. Robert sighs, leans back onto the cot, and listens to the footsteps of the orderly recede down the hallway. He lets go of his notebook, and it falls to the floor. 'Thud,' he says, and laughs. Then he says it again, but more softly and this time without any laughter. 'Thud.'

Some time around 1970, Stanford University professor David Rosenhan called a few of his friends and pitched a crazy idea to them: *Hey, let's all pretend we're insane and get admitted to mental hospitals. Then we'll find out if the doctors can tell the difference between us and real patients!* This suggestion was probably met with polite silence or diplomatic evasions. *That's an interesting idea, Dave, but I'm kind of busy.* However, Rosenhan pushed. *Come on, it'll be an adventure!* And finally he wore his friends down. After all, they figured, why not? It was the 1970s – everyone was doing crazy things!

Bluffing one's way into a mental hospital wasn't a new idea. Soldiers had been doing it for years to get out of fighting. Army doctors called it malingering. In 1887, the journalist Nellie Bly had faked insanity in order to investigate conditions at the Women's Lunatic Asylum on New York's Blackwell's Island. The results of her undercover work, published in the *New York World*, caused a scandal and led to a grand jury investigation of the asylum. In 1952, the anthropologist William Caudill, posing as a patient, checked himself into the Yale Psychiatric Institute in order to study the social structure of asylums. But what Rosenhan imagined was something slightly different and even more ambitious than these earlier efforts. He wasn't interested in dodging military service, exposing conditions at a particular hospital, or conducting an anthropological study. He wanted to expose the practice of psychiatry itself.

The Anti-Psychiatry Movement

By 1970, popular distrust of the psychiatric profession had been building for some time. The counterculture of the 1960s had championed personal freedom – freedom of expression, freedom from restrictive social roles, freedom to think and act as you pleased as long as you didn't hurt anyone else. But this entire social movement seemed to have passed psychiatrists by. Holed up inside decaying mental hospitals surrounded by high walls and barbed wire, psychiatrists were beginning to resemble throwbacks to some bygone era – such as fifteenth-century Transylvania. Many in the counterculture viewed them with suspicion, believing they were no more than guardians of the status quo, enforcers of 'normal' behaviour. Free your mind? Not if the 'mental police', as author Ken Kesey called them, had anything to say about it.

Kesey's 1962 bestseller *One Flew Over the Cuckoo's Nest* communicated this anti-psychiatric sentiment to a huge audience. His novel told the story of a free-spirited convict, Randle Patrick McMurphy, who faked insanity in order to be transferred from a prison work farm to a mental asylum, believing life would be easier in the asylum. But once there, McMurphy discovered that the patients seemed far saner than the staff and doctors. The novel ended on a tragic note. Doctors crushed McMurphy's rebellious nature by performing a lobotomy on him.

An 'anti-psychiatry movement' also emerged among academics

during the 1960s. Ronald Laing, one of the movement's leaders and a psychiatrist himself, denounced conventional psychiatry as dehumanizing. He argued that mental illness was a label the powers that be used to marginalize dissidents and freethinkers. Psychiatric diagnoses, he suggested, were merely a way for society to pigeonhole and ignore that which it didn't understand. Lock the crazies up and forget about them!

Rosenhan was a professor of psychology and law at Stanford when he attended one of Laing's lectures. He sat in the audience, listening carefully, and while he didn't agree with everything he heard, Laing's arguments did get him thinking. How good *were* psychiatric diagnoses? Were people getting locked away in asylums who didn't deserve to be there? As he considered these questions, the idea for an experiment occurred to him. He imagined himself, a perfectly sane individual, confined to a mental hospital. But what if a label on a chart said that he wasn't sane, that he suffered from a psychiatric condition? Would the doctors figure out that the diagnosis was wrong? Would they spot the difference between him and a real patient? If they didn't – if after days or weeks of observation they still clung to a mistaken diagnosis – wouldn't that imply Laing was right, that something was deeply flawed with the process of psychiatric diagnosis?

The idea ate away at Rosenhan. He couldn't put it out of his mind. Perhaps a touch of mid-life crisis was also getting to him. He was married with two kids and settled into a stable, suburban life. So maybe he was looking for adventure. Whatever the reason, he decided to act on his idea. He didn't bathe for a few days. He stopped brushing his teeth. He made himself appropriately funky looking and slightly frayed around the edges. Then, in February 1969, he walked through the door of a hospital and told the staff on duty that his name was David Lurie and he was hearing voices. The voices said, 'It's empty, nothing inside. It's hollow. It makes an empty noise.' He was admitted immediately. Diagnosis: schizophrenia.

Once inside the hospital, Rosenhan became a model of sanity.

He behaved politely. He washed regularly. He was cooperative. Whenever a doctor asked him about the hallucinations, he assured him he was no longer experiencing them. He waited for his sanity to be recognized, but it didn't happen. The weeks stretched by, and finally Rosenhan was discharged. The schizophrenia label, however, stuck. The doctors merely appended the phrase 'in remission' to it.

Rosenhan felt his suspicions had been confirmed. Psychiatrists had been unable to see past the schizophrenia label and recognize his sanity. To him this suggested there were serious flaws with the process of psychiatric diagnosis, but he knew the test wouldn't convince critics. They would dismiss it as a fluke, and anyway, he had slightly tainted the experiment by informing the chief psychologist at the hospital of his plan beforehand – as an escape route in case anything went wrong. So Rosenhan began to concoct an even grander scheme, a test no one could ignore. He imagined an entire group of 'pseudopatients' showing up at hospitals throughout North America, with no advance warning given to anyone on the inside. It would be a full-frontal assault on the psychiatric community. And that's when he started phoning his friends.

The Pseudopatient Study

Rosenhan assembled a group of eight volunteers – five men and three women – including a paediatrician, a painter, a housewife, and himself. Rosenhan carefully coached them on what they were to do and how they were to behave. When they first showed up at the hospital, they were to complain of a single, specific problem – auditory hallucinations, specifically a disembodied voice that said 'thud', 'empty', or 'hollow'. He chose these words, he said, in order to 'lead an observer to suspect an interesting existential problem'.

Once the volunteers were inside, this symptom was to vanish, and they were to act entirely sane. They were not to take any medication. Rosenhan showed them how to fake swallowing a pill by tucking it under their tongue. He made them all invent pseudonyms.

And then he set them loose. One by one, between 1969 and 1972, they showed up at hospitals on the east and west coasts of the United States.

The pseudopatients later recalled that as they walked up to the hospital doors, they felt a mixture of emotions. They were excited by the sheer audacity of the plan, but they were simultaneously nervous that they wouldn't make it past the entrance interview, that the attending physician would immediately see through the ruse. They also feared what lay ahead if they did make it in. Though several of them had previously worked behind the nurses' station in a mental hospital, none of them, except Rosenhan, had any prior experience as a patient. They had only heard stories about what happened at night and on weekends in such places. None of these stories made them any less apprehensive.

They needn't have worried about getting in. They all breezed through the entrance interview. They were uniformly diagnosed as suffering from schizophrenia, except for one of them whose identical symptoms were given a label of 'manic-depressive psychosis'. Once inside, they were led to a room where a doctor gave them a physical exam: *stick out your tongue, bend over, pull down your pants, cough.* Nurses and candy stripers wandered in and out of the room as this went on, seemingly oblivious to the patient with their pants down around their ankles. It was the first lesson in hospital life for Rosenhan's team: they were now just patients; they had no right to privacy.

Day-to-day life on the psychiatric ward turned out not to be as scary as they had feared. In fact, the biggest problem was boredom. There was nothing to do. Patients spent most of their time loitering in the dayroom watching television, while the staff sat in a glassed-in space, nicknamed 'the cage'. The two groups rarely interacted except when the staff ventured out to give everyone, seemingly indiscriminately, copious amounts of pills: Elavil, Stelazine, Compazine, Thorazine, etc. The pseudopatients secreted their pills to the bathroom, only to find the pills of other patients already lying at the bottom of the toilet bowl.

To occupy themselves, the pseudopatients paced the hallways, tried to strike up conversations with those around them, or wrote down observations in a notebook. This latter behaviour quickly aroused the suspicions of the other patients. One time, as Rosenhan sat taking notes, a patient shuffled up and leaned close to him conspiratorially. 'You're not crazy,' he said. 'You're a journalist, or a professor. You're checking up on the hospital.'

'I was sick before I came here,' Rosenhan insisted. 'But now I'm feeling much better.'

All the other pseudopatients had similar experiences. In some cases, they also noticed that the other patients began to imitate them, diligently jotting down cryptic phrases in their own note-books.

However, the behaviour of the pseudopatients didn't seem to arouse any suspicions among the doctors and staff. Quite the con-trary: the constant note-taking was interpreted as a sign of mental disturbance. 'Patient engages in writing behavior,' one nurse wrote in a pseudopatient's record, as if this was cause for some concern.

The pseudopatients had hoped to be out in a day or two, but the days crept by without their release, and as time dragged on, a sense of powerlessness crept over them. The doctors and staff, Rosenhan began to suspect, didn't regard patients as fully human. Instead, they treated patients as if they were animals whose opinion was of no consequence. For instance, he noted that doctors avoided making eye contact with patients and ignored their questions. Frequently the staff would talk openly among themselves about a case, oblivious to the presence of the patient they were discussing, who was sitting right there in the dayroom, hearing everything. Rosenhan described one scene that particularly stuck in his mind: 'A nurse unbuttoned her uniform to adjust her brassiere in the presence of an entire ward of viewing men. One did not have the sense that she was being seductive. Rather, she didn't notice us.'

The pseudopatients also experienced the more brutal side of asylum life. The staff often screamed and cursed at patients. One pseudopatient was woken every morning by cries of, 'Come on, you motherfuckers, out of bed!' Attendants savagely beat patients for minor misdemeanours. To Rosenhan the punishment seemed 'so excessive that it could not be justified by the most radical interpretations of psychiatric canon'. He saw one patient beaten merely for approaching an attendant and saying, 'I like you.' The pseudopatients themselves were never disciplined, but then they made sure to be model patients, 'paragons of cooperation'. Nevertheless, their sense of vulnerability unnerved them. The threat of physical violence hung in the air, held at bay only by the whim of the staff. Rosenhan kept a lawyer on call throughout the experiment in case a pseudopatient needed to be removed rapidly from a hospital, for any reason.

Then one day, out of the blue, a doctor called them in and told them they were free to go. There was no apparent logic to the decision. Why now and not earlier? How had their behaviour changed? The discharge notes offered no clarification. Nor did the notes suggest there had been any suspicion about their status as real patients. As had happened to Rosenhan before, the phrase 'in remission' was simply added to the initial diagnosis. The total length of confinement varied from seven to fifty-two days. The average stay was nineteen days.

A Sword Plunged into the Heart of Psychiatry

Even before Rosenhan officially published his results, word began to spread around the psychiatric community about the experiment. One hospital contacted him, insisting that the pseudopatients wouldn't have fooled them. They could have spotted the fakes! Rosenhan called their bluff. He told them that during the next three months, one or more pseudopatients would attempt to gain entry to their hospital. All they had to do was spot the interlopers. The staff and doctors at the hospital swung into a state of heightened alert, casting a sceptical eye on every person who walked through their door. During the three-month period, 193 people sought admittance. The staff flagged 41 of them as fakes. The doctors, on the other hand, picked out 23 suspect individuals. The staff and doctors agreed on 19 cases. In actual fact, Rosenhan hadn't sent any pseudopatients. He'd been planning to, but his volunteer got sick. *Psychiatric community: 0; Rosenhan: 2.*

Rosenhan published his results in January 1973 in *Science*, one of the most prestigious scientific journals in the world. In the words of Robert Spitzer, a prominent American psychiatrist, it was like 'a sword plunged into the heart of psychiatry'. The sense of outrage was palpable. Criticism poured in, denouncing the study as 'seriously flawed by methodological inadequacies' and 'pseudoscience presented as science'.

Many detractors questioned what the hospital physicians could have done differently. Should they have accused the pseudopatients of lying? But on what basis? Surely it was the right thing to place a person hearing voices under observation.

Rosenhan responded that his critics were misconstruing his argument. The problem wasn't that the hospitals had admitted the pseudopatients. 'If there were beds,' Rosenhan wrote, 'admitting the pseudopatients was the only humane thing to do.' The problem, he insisted, was the initial diagnosis of schizophrenia, which then became a permanent label. Was that diagnosis justified? Why hadn't the admitting physicians described the patients as suffering from auditory hallucinations, and left it at that? Why did they take the extra step of asserting that schizophrenia, with all its negative connotations, was the cause of their problems? It was as if someone were to go to their family doctor complaining of a cough, and immediately be told, without any tests conducted, that they had tuberculosis. Rosenhan argued that schizophrenia had become a 'wastebasket' diagnosis – a vague, catch-all category applied as a label to just about any mental problem.

From a historical point of view, Rosenhan won the argument. After his study came out, during the 1970s and 1980s the use of broad diagnostic categories such as schizophrenia declined sharply. In an effort to be more objective and uniform in their diagnoses, psychiatrists came up with hundreds of new, more specific disease categories, and then systematized their use via a checklist model of diagnosis. *Does a patient have symptoms X, Y, and Z? Then she has syndrome W.*

Rosenhan, however, was only marginally responsible for this change. Insurance companies were certainly the more persuasive catalyst of reform, since they had begun to complain about paying for the treatment of vague psychiatric conditions that never seemed to get better. But Rosenhan was nevertheless on the winning side of the debate.

However, even if Rosenhan's critique of diagnostic procedures did turn out to be prescient, his critics also had a point. From a scientific perspective, his pseudopatient study, for all the attention it got, *was* very strange and not particularly rigorous. It lacked a control group. It was open to charges of experimental bias (his pseudopatients might have unintentionally acted crazy to get the reaction they wanted). It relied heavily on anecdotal, possibly cherry-picked evidence. Rosenhan also provided few details about the size and character of the hospitals involved, making it hard to know how representative they were. Could one really extrapolate from these few examples to psychiatry as a whole?

But then again, maybe these criticisms miss the point. Rosenhan's experiment wasn't any more objective or unbiased than his (fake) patients were schizophrenic. But it wasn't his intention to create an airtight, logically rigorous study. Instead, he wanted to shake things up a bit. He wanted to give the psychiatric old guard a kick in the pants, and he certainly succeeded at that. The blow landed right on target. Thud!

The Seductive Dr Fox

Lake Tahoe, California – 1972. Michael Fox watches as several people walk into the room and take a seat. Although he's been acting for years, he has butterflies in his stomach in anticipation of what he's about to do.

'At least I scored a free trip to Lake Tahoe out of this,' he thinks, 'But there's no way this is going to work.'

He leans over and whispers in the ear of the researcher sitting beside him, 'Someone is going to recognize me.'

'Relax, Michael. Everything is going to be fine.'

'A few words out of my mouth, and they'll know I'm a phoney.'

'Relax!' the researcher repeats, more emphatically.

Another person walks in and takes a seat.

The researcher looks at his watch. 'OK, we should get this started. Are you ready?'

Michael shrugs. 'Ready as I'll ever be.'

The researcher stands up and turns to address the audience. 'I want to thank everyone for coming to today's presentation. It's my pleasure to introduce our distinguished speaker from the Albert Einstein College of Medicine. He's an authority on the application of mathematics to human behaviour. Many of you, I'm sure, are familiar with his work. He'll be talking to us about 'Mathematical Game Theory as Applied to Physician Education'. Please give a warm welcome for the real McCoy, Dr Myron L. Fox.'

Michael rises from his seat, to the accompaniment of a polite round of applause, and walks to the front of the room. He looks out over the audience – eleven faces raised expectantly, waiting to hear whatever wisdom he has to impart. Then he flashes a devil-may-care grin and starts to speak.

In 1972, Michael Fox was a character actor in Hollywood, best known at the time for a recurring role as an autopsy surgeon on *Perry Mason*. He was no relation of Michael J. Fox, of *Back to the Future* fame, who was only eleven years old at the time. However, he was the reason the younger actor later added 'J' as a middle initial, due to Screen Actors Guild regulations that forbade duplicate name registrations.

Fox was between jobs when a trio of Southern California professors – John Ware, Frank Donnelly, and Donald Naftulin –

approached him with an unusual proposal. They wanted to hire him to pose as an academic, the fictitious Dr Myron L. Fox, and give a one-hour lecture at a University of Southern California continuing-education conference to be attended by psychiatrists, psychologists, and social workers.

The topic of the lecture would be a subject he knew nothing about, and that was the point. The professors were sceptical of the value of student ratings of teachers. They had begun to suspect that students rated teachers on their personality, and not on their effectiveness as educators. So they decided to find out how an audience would react to an actor who was charismatic and sounded authoritative, but who was simultaneously extremely uninformative. Would Fox be able to seduce the crowd into believing he was educating them, even if he made absolutely no sense?

Entertaining Nonsense

To craft their deception, the professors started with an article published in *Scientific American* ten years earlier, 'The Use and Misuse of Game Theory'. It was a complex article, but it was nevertheless addressed to an audience of non-specialists. Then they adapted it to their purpose. They preserved the jargon, but stripped it of meaning, coaching Fox to talk about the subject employing 'an excessive use of double talk, neologisms, non sequiturs, and contradictory statements' – while simultaneously mixing in a heavy dose of parenthetical humor. When they were quite sure Fox had mastered the art of being 'irrelevant, conflicting, and meaningless', they set him loose on the crowd.

Fox worried about whether he would be able to pull it off, but he needn't have. For an hour, he dazzled the audience with his wit and humor. They laughed at his jokes and nodded sagely at his explanations. In a masterpiece of improvisation, he then fielded questions for a half-hour.

Fox gave a live presentation before an audience of eleven

people, and subsequently the professors showed a videotape of his lecture to several more groups. All told, fifty-five people watched Dr Fox perform. Immediately after each lecture, the professors asked the audience to fill out an eight-question survey to gauge their reaction. The bad news was that 50 per cent of the audience said Dr Fox dwelled upon the obvious. The good news was that 90 per cent said he presented the material in a well-organized form, and 100 per cent said he stimulated their thinking. In the comments section one person wrote, 'Excellent presentation, enjoyed listening. Has warm manner. Good flow. Seems enthusiastic.' Another person, oddly, claimed to have read some of Dr Fox's publications. Not a single audience member noted that the content of Dr Fox's presentation was pure crap – the term the researchers used to describe it.

Why did the audience respond so positively to Dr Fox? The researchers attributed it to the Halo Effect. Edward Thorndike was the first psychologist to describe this phenomenon, back in 1920. He found that people have difficulty rating others on specific characteristics. Instead, we tend to form a general impression of a person, and this impression, like a halo, influences all our observations of him or her. So, for example, if there's a colleague we dislike, we criticize everything he does. Or if there's a professor whom we think is a nice guy, we overlook the fact that he knows nothing about the subject he's teaching. Naftulin, Ware and Donnelly gloomily concluded: 'Student satisfaction with learning may represent little more than the illusion of having learned.' They recommended – only somewhat tongue-in-cheek – that trained actors should replace professors.

The audience members, for their part, expressed bewilderment when they eventually learned of the deception. Many of them insisted that even if Dr Fox had been a phoney, he had nevertheless stimulated their interest in the subject, and to prove this they wrote to the researchers requesting copies of the original *Scientific American* article from which the lecture had been developed.

Following Dr Fox

The Dr Fox study quickly became something of a cult classic – one of those experiments that a lot of people heard about, though not many read the original write-up since it was published in a relatively obscure publication, *The Journal of Medical Education*. References to it pop up in books about business management and how to give effective presentations. Hang out in any faculty lounge and it's almost inevitable someone will mention it if the subject of student evaluations comes up: *Did you hear the one about that fake professor whom all the students loved?* A 1974 article in British medical journal the *Lancet*, came up with the term the 'Dr Fox Effect' – the tendency for students to equate being entertained with being educated – and the name stuck.

Despite the fame (or notoriety) the experiment achieved, critics complained that the study suffered from methodological problems. For instance, it lacked an experimental control group. The survey given to the audience was limited in its scope. Sceptics argued that even if it was possible to seduce an audience for a single lecture, wouldn't most students have caught on to the ruse if they were stuck with Dr Fox for an entire semester?

To address such criticisms, John Ware and Reed Williams persuaded Michael Fox to reprise the role of Myron L. Fox for a series of six videotaped lectures on 'The Biochemistry of Memory'. In three of the lectures Fox laid on the charm, but in the other three he talked in a dull monotone. Ware and Williams found that not only did subjects who watched the 'high seduction' performance give Dr Fox higher ratings, but they also scored better on a quiz about the material. The Dr Fox Effect appeared to be a robust phenomenon. This implies, rather paradoxically, that if you have the choice between a funny but dumb teacher and a smart but boring one, you should go with the funny-dumb one. You'll learn more. To put that another way, knowledge is useless without a little showmanship to communicate it.

Michael Fox went on to have a long and successful career as an actor. He played the role of Saul Feinberg for eight years on the CBS soap opera *The Bold and the Beautiful* and made appearances on TV shows such as *ER* and *NYPD Blue*. So he avoided being typecast as Dr Fox. However, it's likely that fifty years from now Myron L. Fox will nevertheless be his best-remembered character, even though he never listed it on his official résumé.

CHAPTER FOUR

Monkeying Around

A long time ago, two groups of apes separated in a forest. One of the groups, for reasons not entirely clear, stayed where they were. The other group wandered out of the forest onto the savannah. Some members of this group continued to wander further and further, until they had travelled across the entire globe. Millions of years passed. Descendants of those exploratory apes, who by now looked quite a bit different (they had lost much of their hair and walked upright) returned to the forest and encountered the group of apes from which their ancestors had separated so long ago. 'That's strange,' the hairless, exploratory apes thought. 'These forest apes look a lot like us. I wonder if they're like us in other ways?' Then another idea popped into the minds of the hairless apes: 'By studying these forest apes, perhaps we can learn something about ourselves.' From that peculiar premise, the science of primatology began.

The Man Who Talked to Monkeys

South Shore of the Fernan Vaz Lagoon, West Africa – April 1893. Richard Lynch Garner, a burly white man with a thick, well-groomed moustache, sits in a cage. He stares out through metal wire at the wall of vegetation surrounding him. The air is hot and humid, and it carries the low hum of insect life. He peers deep into the shadows, seeking out any movement, any sign he isn't alone. Seeing nothing, he returns his attention to the notebook open on the table in front of him. He picks up a pen, puts it down, then picks it up again and begins to write:

Here I sit, in a stillness almost as great as that of a tomb . . . unspeakable silence . . . While it is true that this great forest teems with life, there are times when it appears to be an endless, voiceless solitude . . .

Every breeze is laden with the effluvia of decaying plants. Every leaf exhales the odours of death.

Garner stops writing and pushes the notebook away. The table and chair are the only furniture occupying the cage with him. To his left is a stack of canned meat, and on his right, within easy reach, lies a rifle. Leaning back in his chair, he waits.

An hour passes, and he cocks his head, as if he's heard something. A twig snaps, confirming his suspicion. He jerks around, only to see a fat armadillo stroll lazily into the clearing. It sizes him up with its beady eyes, but sees nothing about the caged human to be concerned about. It sniffs the ground nonchalantly and continues back into the jungle. Garner sighs and settles back into his chair.

More hours pass, and the equatorial heat reaches its midday peak. Even the constant insect hum recedes beneath the relentless onslaught of the sun. All is still. Garner mops his brow with a handkerchief, takes a drink of water from a canteen, and continues to wait.

Eventually the cool of late afternoon arrives, and the jungle returns to life. Far away in the distance an animal screeches. It's answered, even further away, by a howl.

And then Garner tenses. He's heard a rustling of leaves. He looks back and forth, but it's a minute before he locates a pair of dark-brown eyes watching him from the shadows. Around the eyes he can barely discern the shape of a gorilla's head.

Immediately, he stands up and approaches the wire of the cage. He can sense the massive bulk of the gorilla, partially concealed by branches.

He can feel its strength. The metal wire protecting him suddenly seems thin and fragile.

The gorilla doesn't move as Garner returns his stare. The gaze continues for several moments. A tense silence hangs between the two primates, and then Garner tilts his head back and utters a soft, low-voiced cry.

'Waa-hooa . . . Waa-hooa.'

Startled, the gorilla takes a step back into the jungle. It hesitates and then stands up, fully revealing its face.

Garner repeats the call. 'Waa-hooa . . . Waa-hooa.'

He clutches his fingers around the metal wire. His pulse races, and sweat drips down his neck. His natural impulse is to look away, to retreat from the threat of the larger animal, but he stands his ground, waiting to see if he'll get a response. The gorilla opens its mouth, as if about to say something. Garner leans closer. But then the creature changes its mind. It utters a loud grunt of disinterest, 'Umph,' turns, and silently disappears back into the jungle.

Nine years earlier, Richard Garner could have been found standing and staring through the bars of a different cage, but back then, in 1884, the situation was reversed. Garner stood on the outside of the cage, jacket draped over his arm and sleeves rolled up, intently watching a group of primates locked inside – a troop of small, brown monkeys and one large, long-muzzled mandrill.

By trade, Garner was a businessman. He sold stock and real estate, a job that required him to criss-cross the country to meet his customers. Finding himself in Cincinnati, he'd decided to spend the day at the zoo that had recently opened there in 1875. It was the second zoo to open in the United States, and like most Americans at the time, Garner had never visited one before.

Crowds of people moved around Garner. Children, excited at a day out, ran past clutching bags of roasted peanuts. Some paused to laugh at the monkeys before hurrying on to the bear or lion cages. Young couples strolled by hand in hand. Garner was oblivious to all of them. He was aware of nothing but the monkeys.

The monkey enclosure consisted of several large cages, separated by a brick wall. A small door allowed the monkeys to come and go between the cages as they pleased, but despite this freedom of movement, all of them had congregated in a single cage, nervously watching the mandrill that squatted in the adjoining quarters. Every movement of the mandrill set off a furious chatter among the monkeys.

Garner had been watching the monkeys for hours, and the more he observed them, the more it seemed to him that their chatter was not random. He felt sure it had meaning – that the monkeys with the best view of the mandrill were informing their fellow troop members, with specific cries, of every movement of the threatening creature. *Ereck-ereck: Look out, the mandrill is standing up! Whoo-whoo: All is safe. The mandrill is lying down.*

Garner continued to stand in front of the cage for hours that day, and he returned the next day, and the day after that. With each visit he grew more convinced that the monkeys were using a primitive form of language to communicate with each other. In fact, if he closed his eyes he believed he could tell, merely by the sounds of the monkeys, exactly what the mandrill was doing.

Garner was thirty-six years old and lacked any formal scientific training. He'd drifted through life up to that point, first training to become a minister before abandoning that to become a schoolteacher and then a businessman, but as he stared at the chattering monkeys he knew he had, at last, found his true calling. He resolved to make it his mission to break through the language barrier that separated man from the other primates and become the first human to master the 'simian tongue'. He resolved to become the first man to converse with monkeys.

Learning the Simian Tongue

From that day in 1884 onward, Garner embarked on a self-guided monkey-language research program. Wherever he found monkeys, he spent long hours patiently observing them, making notes about their behaviour and vocalizations. He examined monkeys in travelling shows, accompanying hand-organ players, and kept as family pets. He became a familiar figure at the zoos of New York, Philadelphia, Cincinnati, Washington, and Chicago. At first the zookeepers dismissed him as a kook when he marched up and announced, 'I want to hear your monkeys talk,' but gradually they accepted him, even if they continued to regard him as something of an eccentric.

Garner's greatest frustration was the problem of accurately transcribing monkey sounds onto paper so that he could compare and analyse them in greater depth. Many of the sounds simply defied transcription. For instance, he came up with 500 variations of 'egek' in an effort to get it right. Twenty years earlier, a British naturalist, Colonel Samuel Richard Tickell, encountered a similar problem when he tried to transcribe the melodic mating calls of the white-handed gibbon of south-west Asia. The solution Tickell came up with, which he described in an article published in the *Journal of the Asiatic Society of Bengal*, was to render the calls into musical notation. However, Garner was probably unaware of Tickell's work, and even if he'd known of it, it wouldn't have helped him much. The raucous chatter of rhesus and capuchin monkeys wasn't amenable to musical notation.

Another problem was getting the monkeys to repeat their calls on command. The monkeys, as if trying to be difficult, clammed up when Garner took out his notepad, or they refused to make the sound Garner was interested in. The work was tiring and thankless. He later confessed he was on the verge of 'being shipwrecked' when, out of the blue, a solution to his problems presented itself.

In storefronts, during the late 1880s, a new invention had recently appeared. Crowds gathered to marvel at the strange contraption that spoke words from its mechanical innards. *I am the Edison phonograph*, the machine announced as part of its prepared advertising pitch, *created by the great wizard of the New World to delight those who would have melody or be amused. . . . If you sing or talk to me, I will retain your songs or words, and repeat them to you at your pleasure. I talk in every language. I can help you to learn other languages.*

The idea came to Garner in a flash – *use a phonograph to record monkey sounds!* He was so excited when he thought of it that for many nights afterwards he could barely sleep. He lay awake tossing and turning, thinking of all the possibilities. Finally he scraped together his money and bought one of Edison's machines, and he dove into his monkey-language studies with renewed vigour.

Garner's first step was to test the reaction of one monkey to a recording of another. He described the concept to Frank Baker, an anatomist at the Smithsonian Institution, who agreed it would be an interesting experiment and granted Garner access to a pair of male and female monkeys at the Washington zoo. Garner separated the pair and persuaded the female to make a few sounds into the phonograph. He then played this recording to the male. The monkey's reaction was instantaneous. It stared at the machine, per-plexed. Then it approached the phonograph and looked behind it. It looked all around. It stared questioningly at the horn. Finally, it thrust its arm into the instrument, all the way up to its shoulder, withdrew it, and then peered in again. Garner judged the experi-ment a complete success. Clearly the monkey had recognized the sound.

Garner carried his phonograph to other zoos, where his high-tech experimentation soon brought him to the attention of journalists. However, they had a hard time taking his research seri-ously. He seemed a slightly absurd figure as he chased monkeys around their cages, trying to convince the small creatures to speak into the recording trumpet. The reporters suspected Garner was a bit mad, but at least, they figured, his exploits were entertaining in a comical way. They nicknamed him 'Monkey Man'.

It was far from easy to record the monkeys because they tended to clam up whenever Garner approached with the phonograph. A *New York Herald* reporter who observed Garner's efforts noted, 'The little scamps utter their cries freely enough, but the shrewdest of lawyers could not guard himself more carefully against going on record.' But through his persistence, Garner gradually amassed a library of their sounds. This allowed him to progress with the second stage of his plan, which was to learn to imitate their cries and test how monkeys would react to his vocalizations. He envisioned becoming a fully-fledged 'simian linguist'. Night after night, when all else was silent, he listened to the cylinder recordings and repeated what he heard. People walking by outside must have been puzzled to hear the cries and counter-cries of monkeys coming from his room.

Finally Garner felt ready to test his verbal skills on a live subject. He approached a small brown capuchin monkey named Jokes housed in a South Carolina zoo. While Jokes was eating from his hand, Garner spoke a sound that he believed meant 'alarm' or 'assault'. Alphabetic letters, Garner said, couldn't accurately represent the sound, but he described it as resembling a short 'I', 'uttered in a pitch about two octaves above a human female voice'. As soon as the cry came out of Garner's mouth, Jokes leapt backwards onto the top perch of his cage. He stared at Garner, his eyes wide with fear. Despite Garner's most conciliatory efforts, Jokes refused to come down from that position. Days later, he was still eyeing Garner with suspicion.

The positive result from his attempt at monkey communication pleased Garner, but he was restless to achieve more. Experiments with monkeys in zoos were all well and good, but what might he learn, he wondered, if he travelled to Africa and studied the great apes in their natural setting? Once the idea came into his head, he couldn't shake it. He decided he had to do it. He would go to Africa! Letting his imagination run wild, he foresaw not only communicating with chimpanzees and gorillas – whom he assumed had far greater linguistic powers than monkeys – but also opening up trade relations with them. 'One very important experiment that I contemplate trying,' he wrote, 'is to see if I cannot effect a very limited commercial treaty. I believe it is possible to exchange certain articles with those apes, which will indicate in them a feeble appreciation of values and the law of exchange.' Unfortunately he never elaborated on what goods exactly he intended to trade with the apes.

Although Garner's ambitions were large, his finances were meagre, so he had to appeal to the public for investments and donations. He tirelessly pitched his expedition to whomever was willing to listen, painting it as a high-tech adventure in the service of science – a kind of Jules Verne-style excursion outfitted with the latest gadgets and equipment. Jules Verne, fittingly, later wrote a novel inspired by Garner's exploits, *Le Village aérien* (*The Village in the Treetops*). Certainly nothing quite like Garner's expedition had ever been attempted before. He planned to take a phonograph into the jungle to record the apes, but not just any phonograph. It would be a special, double-spindle design, custom built for him (he hoped) by Edison's engineers, allowing him to record and play sounds simultaneously. Telephone receivers placed throughout the jungle would transmit sounds back to the phonograph, thereby greatly expanding the area he would be able to monitor. Cameras with bait-activated triggers would surround his camp. To ensure his safety, Garner planned to sit inside a large steel cage, electrified to deter predators. Finally, he would take an array of weapons – guns, steel-tipped arrows, and ammonia-filled canisters – in case of attack.

The idea received a substantial amount of press, but money was slow to come in. By early 1892, Garner had only raised $2,000. He had hoped for $10,000. Part of the problem may have been that people still had a hard time taking him seriously. Learning ape language just didn't sound like real science. He began to fear he wouldn't be able to go. Instead he sat down and revamped the expedition, abandoning details such as the jungle telephones and trigger-activated cameras. It was no longer going to be the grand safari he had envisaged, but he decided he could manage. So he bought his provisions, booked his passage, and set sail from New York City on 9 July 1892, bound for the jungles of Africa.

Into the Jungle

Up to this point, the scientific community had paid Garner little attention. Like everyone else, they dismissed him as a well-meaning, if slightly misguided, amateur. But when Garner announced he was departing for Africa, it became harder to ignore him. Garner's expedition, eccentric though it might be, was nevertheless going to be the very first scientific study of chimpanzees and gorillas in the wild. And what if he actually found a talking ape? The few scientists who knew anything about apes didn't think this was likely, but so little was known about the creatures that they had to admit it was possible. If Garner turned out to be right, it would be a remarkable finding that would shatter the widely held belief that humans alone possessed language.

Garner himself fully anticipated he was going to return from the jungles of Africa with evidence of a linguistic 'missing link'. In his daydreams, he imagined he would be hailed as the next Charles Darwin, his personal hero. He fantasized about then lording this triumph over all those who had doubted him, particularly the professional scientists whom, he resentfully felt, should have been more receptive to his ideas.

However, the scientific community needn't have worried about being upstaged. Garner had a knack for ineptitude, thanks to which things started to go wrong for him even before he got to Africa. For some inexplicable reason, he didn't purchase the phonograph – the most important piece of equipment for his expedition – while he was still in the United States. Instead, he figured he could pick one up in England, where he planned to stop en route to Africa. But when he showed up at the British branch of Edison's company, the salesmen there, although they'd been warned of his arrival, brushed him off by demanding an outrageous price for the machine – a price Garner couldn't afford. He cursed and raged. He screamed that they were mindless bureaucrats, but they wouldn't back down. Garner had unwittingly got caught up in the politics of international business. Edison's British representatives didn't actually see it as their job to sell phonographs. Instead, their primary concern was to enforce Edison's patents and to prevent rival European companies from selling the machines. To them, Garner with his bizarre request was just a nuisance. Reluctantly, he left empty-handed.

Garner blamed his failure to get a phonograph entirely on Edison's representatives, but his own poor planning was certainly equally to blame. Edison, in fact, had been supportive of Garner's research. The famous inventor later pointed out that Garner had a standing invitation to come to Menlo Park and work with his engineers on designing the custom-built phonograph, but Garner never took Edison up on the offer. Perhaps he didn't have the time or money to make the trip. Whatever the reason, the lack of a phonograph proved to be a serious setback.

Nevertheless, Garner refused to admit defeat. In September 1892, he loaded his man-cage onto a ship at the quayside in Liverpool as a crowd looked on. He waved goodbye to everyone, boarded the ship, and continued on his journey. A month later, he arrived on the coast of Gabon, where he transferred his equipment onto a small steamer and travelled 250 miles up the Ogowe River, before cutting across country to the south shore of the Fernan Vaz Lagoon, in the territory of the Nkami tribe. There, in April 1893, he erected his cage, christened it Fort Gorilla, and settled in for what he expected to be his 'long and solitary vigil' in the jungle.

The first thing Garner discovered about Africa, as he sat in his cage, was that it wasn't as solitary a place as he had imagined. It was actually full of people, including many Europeans. The problem was how to avoid them. There was a French mission a mile from his camp, close enough for him to hear the bells from the small chapel, and the locals often walked out to stare at the curious sight of a man locked in a cage. For them it was like some kind of reverse zoo – the white explorer trapped behind bars as the animals roamed free in the jungle around him.

The next thing he discovered was that Africa was full of mosquitoes and foreign diseases. He spent two months feverishly ill.

Finally, as he sat waiting for wild chimpanzees and gorillas to show up, he realized there was a serious flaw in the design of his study. Remaining in one place, hoping the apes would come to him, turned out not to be a good way to make contact with such cautious creatures. He was deep in ape country, and there were plenty of them around, but sensibly they avoided the disturbing presence of a man sitting in an electrified cage with a gun.

Sometimes, as Garner lay in his cage at night trying to sleep, he thought he heard the calls of gorillas far away in the jungle: *Waa-hooa*. *Waa-hooa*, answered by *Ahoo-Ahoo*. He imagined this meant, 'Where are you?' and 'I'm here!' Occasionally gorillas even meandered into view on their way to find fruit in the nearby plantain grove. One time he heard something creeping through the bushes and looked up to see a gorilla seven yards from his cage, holding onto a bush with one hand, its mouth open as if in surprise. Garner froze, afraid to move lest he frighten his visitor away. The gorilla stared at him for several moments, uttered a loud 'Umph,' and then wandered off. This was as close to contact as Garner came.

To pass the time, Garner purchased a young chimpanzee whom he named Moses. He kept him in a small pen next to his cage, and during the long days tried to teach him to speak. This wasn't what he had travelled all the way to Africa for – he could have studied a captive chimpanzee in a zoo back in the United States – but it kept him busy. He reported that he managed to get Moses to say *'Feu,'* the French word for fire. He theorized, probably correctly, that this was the first human word ever spoken by an ape.

Nine months passed in the jungle. Moses contracted an illness and died. Sadly, Garner buried his little friend, but by this time, he was out of money anyway. So he packed up Fort Gorilla and began the long journey home. His grand expedition to find and record talking apes had been a failure.

The Quest Continues

Garner arrived back in the United States in late March 1894. His return made front-page news, but it wasn't the hero's welcome he had been hoping for. Instead, reporters pestered him with annoying questions: *What happened to the phonograph? Where were the talking apes?* Garner didn't have an exciting story to tell the papers, so instead, to entertain their readers, journalists invented bizarre, tongue-in-cheek rumours about his expedition. For instance, it was reported that Garner had hypnotized a chimpanzee with a mirror and found an orangutan that could sing 'Tonner and Blitzen'. Garner realized that the scientific acclaim he had hoped for was as distant as ever. He was still the butt of jokes, still the eccentric Monkey Man.

A different man might have given up at this point and gone back to selling stock and real estate, but that wasn't Garner's way. He was nothing if not stubborn. Instead he redoubled his commitment to his life's work. The idea of finding a talking ape became his obsession, and he resolved to do whatever it took to get back to Africa in order to continue his research. After casting about aimlessly for a while, he discovered there was good money to be made by collecting specimens in Africa for the many natural-history museums that were popping up in America, such as Chicago's Field Museum (founded in 1893), Pittsburgh's Carnegie Museum (founded in 1895), and the American Museum of Natural History (founded in 1869). A gorilla skeleton could fetch up to $200. An African manatee went for $100. That was enough to fund his travels, and this became his new vocation.

Garner spent most of the rest of his life in Africa, searching for butterflies, birds, monkeys, and whatever other specimens he could sell to museums. But like Captain Ahab hunting down the white whale, he constantly remained vigilant for any signs of a talking ape, though the elusive creature always remained just out of his reach.

Years passed by, and Garner grew old. It looked like he would never find what he was searching for. But then, in 1919, the *New York Times* reported that Garner, at the age of seventy-one, had at last fulfilled his life's ambition. He'd found a talking ape. The article claimed it happened while the explorer was following a tip from African natives about the existence of a curious group of creatures that appeared to be gorilla–chimp hybrids. Under cover of darkness, Garner crept up close to the creatures' habitat and eavesdropped on them. They sounded like a group of men sitting around conversing.

A few nights later, Garner crept back. He heard one of them calling to his mate: *Waa-hooa . . . Waa-hooa*. Trembling with anticipation, Garner stood up and called back: *Ahoo! Ahoo!* At first, there was no response, so he tried again. *Ahoo! Ahoo!* Then a reply: *AAAHHHOOOOO!* Garner's heart leapt with joy. He had, at last, conversed with a gorilla. But seconds later he realized his mistake. The male gorilla, following what it thought was the call of its mate, came crashing through the jungle towards him. It grew closer and closer. Garner had no protection. The gorilla was only yards away. Fearing for his life, Garner acted on instinct. He lifted his rifle to his shoulder and fired. The gorilla fell to the ground, mortally wounded.

If the tale was true, it represented a strange and tragic conclusion to the explorer's career. He searched his entire life for a talking ape, finally found one, and shot it dead. But then again, it's just as likely the story was a reporter's idea of a joke.

Garner returned to the United States soon after the encounter with the talking ape was reported. He didn't comment on the story publicly. The next year, while in Chattanooga, Tennessee, he suddenly took ill in his hotel room and was rushed to a hospital, where he died of pneumonia. But even in death, the recognition he yearned for eluded him. For several days his body lay in the morgue, unidentified. Finally, while going through his possessions, the hospital staff realized who he was – Richard Garner, the famous Monkey Man! And still the jokes continued. The next year, a syndicated columnist, perhaps not realizing he had died, asked rhetorically, 'Whatever became of that Prof. Richard Garner who went to Africa to study monkey language?' The question was a setup for a one-liner. 'He's probably camped down in New York,' the reporter wisecracked, 'among the cake eaters and flappers trying to decipher their talk.'

Garner died when primatology was on the cusp of becoming an important, prestigious science. The idea of communicating with apes by teaching them human language would be a central concern of this new science, but Garner's notion that apes and monkeys already had a language of their own was treated as something of a joke, as was Garner himself. He was written out of the discipline's history.

The few monkey-language experiments conducted during the following decades were more like publicity stunts than serious undertakings. For instance, in 1932 a Polish researcher, Tadeusz Hanas, dressed up in the skin of an ape and took up residence in the primate cage of the Warsaw Zoo, declaring that, by this means, he hoped to learn the habits and language of apes. When he left his primate companions eight weeks later, he reported that their language was quite similar to that of men.

Five years later, the staff of Washington's National Zoo made recordings of the 'love calls' of Suzy the orangutan. The zoo's director, William Mann, took these recordings to Sumatra, hopeful that a suitor would wander out of the jungle in response to her winsome cries. Unfortunately the wax records were damaged en route, so Suzy's mate had to be found by more conventional means.

But monkey language was suddenly taken more seriously when, on 28 November 1980, a curious headline appeared on the front page of the *New York Times*: 'Studies in Africa Find Monkeys Using Rudimentary "Language"'. The accompanying article described the work of a group of Rockefeller University scientists who had studied free-ranging vervet monkeys in Kenya's Amboseli National Park.

While observing the monkeys, the researchers noticed that the primates warned each other of approaching predators with cries of alarm, and that their cries sounded different for each predator, as if they had specific words for leopard, snake, and eagle. The scientists tape-recorded the cries, then played back the recordings to other members of that monkey group. Sure enough, when the monkeys heard the alarm for an eagle (a series of staccato grunts), they ran into the grass. The snake cry (a high-pitched chatter) made them examine the ground around them. The leopard cry (a chirp-like call) provoked a mad scramble into the nearest trees. The *New York Times* noted, 'The scientists view the calls as real messages, something previously believed rare or nonexistent in animal communication.'

No mention was made of Garner. The researchers later admitted they'd never heard of him, which gives one an idea of how completely science had forgotten him. But to anyone familiar with his work, the Amboseli research was like déjà vu. There were the speaking monkeys and recorded warning cries, just like a page out of one of his books.

Subsequent research reinforced the Amboseli findings when Ivory Coast monkeys were observed using similar cries to communicate messages. In other words, Garner's idea of talking monkeys may have been slightly crazy, but it wasn't as completely crazy as everyone thought. The evidence is now compelling that some monkeys do use vocalizations, in a speech-like way, to communicate specific messages.

One imagines that if Garner could hear this news it would cause him to turn restlessly in his grave. If you listen closely, you might hear him muttering, *I was right! I was right! Waa-hooa! Waa-hooa!*

Chimpanzee Butlers and Monkey Maids

Smedes, Mississippi – 1898. 'This better be good, Mangum. I'm sweating like a tusk hog.' Parkes removes a handkerchief from his pocket and mops the sweat off his thick, sunburned neck.

Mangum smiles. 'I don't think you'll be disappointed. We're almost there.'

The two men continue down the dirt path in silence. Soon they arrive at a vine-covered embankment that Mangum bounds up in a few steps. His companion, huffing and puffing, follows behind more slowly, holding on to the vegetation for support.

The top affords an expansive view of a large cotton field that stretches out for several acres on the other side. Heat shimmers in the air above the plants. After Parkes catches his breath, Mangum gestures dramatically at the view. 'So here we are. I present to you the future of farming!'

Parkes blinks quizzically as he looks to his left and right. 'This is what you brought me to see? There's nothing here but a field of cotton.'

'Not just cotton, my friend. Look closer.' Mangum points at the field.

Parkes follows the direction of Mangum's finger and studies the neat rows of cotton, vibrant green in the noonday sun. He's about to shrug his shoulders dismissively, when out of the corner of his eye he spies a small brown shape, about two feet high, dart from beneath a cotton plant. He turns his head to look and then sees another of the brown shapes, and then another. 'What in the world?'

Mangum grins broadly. 'Monkeys!' he declares. 'The cotton in this field is being picked by a troop of twenty trained monkeys.'

Parkes looks at Mangum as if he's crazed.

'Last year I obtained twenty monkeys from a New York trainer,' Mangum explains. 'It took several weeks to train them to pick the cotton, but they've taken to it with enthusiasm.'

As Parkes stares down at the field beneath him, he can now see the monkeys scurrying about. Each animal wears a small white bag slung over its shoulder. They move energetically from plant to plant, picking the cotton and placing it in their bags.

Parkes shakes his head. 'Well, if I wasn't seeing it with my own eyes, I wouldn't have believed it was possible. Is this profitable?'

'Absolutely. The monkeys do the same amount of work as a human for one-third the cost. What's more, they're more careful and pick a cleaner grade of cotton, and they'll pick rain or shine, on days when you couldn't get a person out here for all the gold in California! I tell you, I believe monkeys to be the greatest discovery made for the cotton planter since Whitney discovered the cotton gin.'

According to ancient legends told by the tribes of western Africa, chimpanzees and gorillas have the ability to speak but conceal this ability from humans, fearing men will put them to work if the truth is known. Hundreds of years ago, when the first Europeans to encounter apes heard these legends, they tried to allay the fears of the primates. 'Speak, and I will baptize you!' Cardinal de Polignac guilefully promised an orangutan displayed in a glass cage at the Jardin du Roi during the early eighteenth century. The orangutan didn't take the bait.

Of course, the fears of the apes were entirely justified. As the twentieth century approached, and primates began turning up in greater numbers in Europe and America, the human race revealed its hand. Entrepreneurs began hatching plans in which they envisioned the profitable conscription of apes and monkeys into the workforce.

Monkeys Pick Cotton

One of the earliest plans to exploit monkey labour was detailed in the American press in 1899. It was a bizarre report about an attempt to train monkeys as agricultural workers in cotton fields. Picking cotton was slow, expensive work – it wasn't until well into the twentieth century that inventors figured out a way to mechanize the process – so the idea of replacing human workers with potentially cheaper, faster monkey labour had obvious appeal. The report originated in a letter printed in the *Cotton Planters Journal*, a respectable trade publication. The letter, from an unnamed correspondent, claimed that the wealthy Mississippi planter W. W. Mangum, with the help of the botanist Samuel Mills Tracey, had successfully trained monkeys of the '*Sphagtalis vulgaris*' variety to work in his fields. The correspondent wrote:

> It was a glorious sight to see, and one that did my heart great good. The rows were filled with monkeys, each with her little cotton sack around her neck, picking quietly, without any rush or confusion. When they got their sacks full they would run to the end of the row, where a man was stationed to empty them into a cotton basket, when they would hurry back to their work. The monkeys seemed actually to enjoy picking.

From the perspective of the twenty-first century, it's tempting to dismiss this report as a farce. After all, *Sphagtalis vulgaris* is a nonexistent species of monkey. However, nineteenth-century readers took it quite seriously, though they acknowledged the claim to be extraordinary. Word of Mangum's achievement travelled as far as New Zealand, where the *Bruce Herald* noted it would have been hard to believe 'if it were not accompanied with names, dates, places, and circumstances, which put deception out of all probability'. Both Mangum and Tracey were well-known persons, and neither of them denied the report in public. So perhaps the two of them did conduct some kind of simian labour experiment down in Mississippi. Although when the *New York Times* reported in 1902 that President Theodore Roosevelt had visited Mangum's plantation while on vacation, it made no mention of any monkey labourers. If they had existed, that would have been a strange omission.

So the story was almost certainly a hoax. But even so, it offered a revealing window onto contemporary attitudes. First, a lot of people thought the monkey workers were a really good idea! Second, it was no coincidence that ideas about primate labour took root in a region that had only recently seen the end of human slavery. To the former slave owners there was a seductive logic to the idea. The slaves were now free, but perhaps their place could be filled by a near-human species that would prove to be both stronger at work and easier to control.

The Monkey with a Mind

As the twentieth century began, scientists undertook the first studies of primate intelligence, and their reports, breathlessly broadcast by newspapers, fanned hopes that teaching apes to serve mankind might be a relatively simple process. For instance, from Africa came frequent updates about the work of the self-proclaimed 'simian linguist' Richard Garner (whom we just met), who claimed to be making steady progress in his efforts to learn the language of monkeys and teach them human speech.

Closer to home, on 9 October 1909, University of Pennsylvania professor Lightner Witmer conducted a highly publicized test of ape intelligence. His subject was a vaudeville chimp named Peter, promoted by his British owners, Joseph and Barbara McArdle, as the 'monkey with a mind'. Of course, Peter was technically not a monkey, but the epithet stuck nevertheless.

On a sellout tour of Europe and North America, Peter stunned audiences with his act in which he smoked a cigarette, sipped tea like a gentleman, and careened wildly around the stage on roller skates and a bicycle. After seeing this show and being highly impressed, Witmer asked the McArdles to bring Peter to his lab, so that he could test his abilities in a scientific fashion.

Witmer arranged for the test to be conducted in a large lecture hall, with a crowd of over one hundred academics in attendance, including the psychologist Edwin Twitmyer, the mathematician George Fisher, and the neurologist Seymour Landlum. Despite such an august gathering, Peter promptly lowered the tone with his grand arrival, bursting into the room on roller skates, dressed in black cloth trousers, waistcoat, tuxedo tails, and silk hat, and then madly skating around the hall with his screaming owners in hot pursuit.

Once Peter had been caught, and order restored, the testing began. Peter handily mastered a series of challenges. He lit a cigarette, threaded beads onto a string, hammered a nail, and put a screw into a board. Language skills came next. 'Say Mamma!' Barbara McArdle ordered him. It took a lot of coaxing. Peter kept wringing his hands in distress, but finally he said, 'Meh . . . meh,' in a loud whisper.

The high point of the day came at the very end, with the writing test. Witmer picked up a piece of chalk and approached a chalkboard. 'Peter, I want you to do this,' he said as he drew the letter W. Peter hadn't been paying attention, so Witmer repeated the command, 'Now look, this is what I want you to do,' and traced a second W over the one he had just drawn. This time Peter focused on Witmer's actions, and he obediently gripped a piece of chalk in his hand and drew a shaky but legible W on the board. Gasps rose from the crowd, followed by enthusiastic applause. Vaudeville tricks were one thing, but writing a letter was something else entirely. That implied a very different, and very human, kind of intelligence.

Witmer spelled out the implications of what Peter had done for the press. 'If this chimpanzee can be taught to make one letter,' he said, 'he can be taught to make other letters on a blackboard. And I believe if he were made to combine certain letters into words, and was shown the object which the word stood for, he could associate the word with the object.' The press responded enthusiastically. The *Oakland Tribune* declared, 'MONKEY COULD BE TAUGHT TO SPELL!'

Witmer was so impressed by Peter that he felt the chimp should become 'the ward of science and be subjected to proper educational influences'. This grand plan, however, never came to pass. Peter continued performing for the McArdles, and he died a few months later, apparently due to exhaustion from his gruelling schedule.

William H. Furness, one of Witmer's colleagues at the University of Pennsylvania, continued the work with apes, attempting to train first an orangutan and then a chimp to speak. He did this by holding their lips and tongue in the appropriate positions. For instance, in order to make the orangutan say, 'Cup,' he first placed a spatula on her tongue, waited until she had taken a breath, and then clamped his fingers over her nose, forcing her to breathe through her mouth. As he did so, he quickly removed the spatula, which produced a 'Kaaa' sound as she exhaled. Next, he quickly pressed her lips together, transforming the 'Kaaa' into a raspberry-like 'Ppppppp.'

After a couple months of this, he concluded, 'I should say that the orang holds out more promise as a conversationalist than does the chimpanzee.' But both of them were a long way from reciting Shakespeare. He reported more success when he lowered his ambitions and trained the chimpanzee to shovel coal.

Gland-Stimulation and Hybridization at the Pasteur Institute

The experiments of Witmer and Furness had a powerful effect on the public mind, encouraging the belief that apes could quite easily be trained to act like men, and the purpose of this transformation, it was widely assumed, was to then make apes do the work of men. A two-part series that ran in the *Washington Post* in 1916 spelled out this ambition explicitly. 'What more dazzling, fascinating project can we conceive,' the author of the article asked, 'than to train this now useless animal loafing in the forests of Africa and Asia to do humanity's most irksome labors?' After summarizing the most recent scientific research on primate intelligence, the reporter rhapsodized about a future in which apes would hoe corn, weed gardens, dig in mines, feed the furnace fires of steamships, and sweep streets, as humans relaxed in their homes, free to pursue more intellectual goals.

The author of the *Washington Post* article was particularly taken by a proposal recently published in the *New York Times* by Dr Jules Goldschmidt of the French Pasteur Institute. Goldschmidt sought American money to help him build a research lab on an island off the coast of Africa where he could conduct experiments on chimpanzees and gorillas. He assured sensitive readers that vivisection was not what he had in mind. Instead, he imagined it might be possible to artificially stimulate the glands of apes in order to 'modify the anthropoid brain' and thereby 'enlarge the infantile cranial cavity, stimulate simultaneously the growth of the brain and obtain a superlative intellectual power'. Ultimately, Goldschmidt suggested, it might be possible to modify an ape brain to such a degree 'that it will approach man's brain as developed among the lower human races'. Goldschmidt promised that he would then apply what he'd learned from his simian research to humans, allowing him to breed geniuses whose intelligence would benefit the entire human race.

By 1923, Goldschmidt's gland-stimulation project hadn't progressed far, but his colleagues at the Pasteur Institute had succeeded in establishing an ape colony in Kindia, French Guinea. News of the colony rekindled hopes that the day of chimpanzee butlers wasn't far off. An article, widely distributed by International Feature Service, Inc., reported the colony was to be a full-fledged finishing college 'to make chimpanzees human'. Chimpanzee pupils – pictured in an illustration wearing tailcoats and striped pants – would supposedly be taught how to talk, read, and write. They would live in a four-storey dormitory, sleep in beds, be served cooked food, attend classes during the day, and after class engage in recreational activities such as listening to phonographs and working out in the gym. 'Results of the most astonishing nature are confidently looked for,' readers were assured.

However, the reality of the Kindia research station was more prosaic. There were no how-to-be-human classes for the captive apes. Instead, the colony was primarily used for tropical disease research – with one significant exception. In 1927, the station opened its doors to the Russian biologist Il'ya Ivanov, who was interested in breeding a human–ape hybrid.

The concept of creating a human–ape hybrid wasn't anything new. European researchers had been toying with the idea for years. As early as 1905, the Pasteur Institute had supported the efforts of the embryologist Hermann Marie Bernelot Moens who tried to organize an expedition to Africa to conduct hybridization experiments. But Moens's plans had fallen apart in the face of public outcry. Now, twenty-two years later, the Pasteur Institute was trying again to bring a 'humanzee' into the world by supporting Ivanov, who conveniently had funding for his research from the Soviet government.

Ivanov, and Moens before him, both argued that the creation of a human–ape hybrid would prove the theory of evolution. Or, at least, it would prove the evolutionary link between humans and apes. But there was a second, more practical reason lurking behind this line of research. As one of Ivanov's American supporters, Howell S. England, told members of the American Association for the Advancement of Science in 1932, hybridization could produce 'working men of more powerful muscle than any now living'.

Similarly, the German paleontologist Gustav Steinmann speculated about the possibility of 'breeding a cross between one of the lowest races of savages and the gorilla' in order to create a creature that could 'do man's rough, hard work'. As reported by the *Pittsburgh Press* in 1924, '[Steinmann] believes that in this way he can produce a creature with the strength of a gorilla and the intelligence of a low type of man. The savage races, as a rule, are disinclined to hard labor, but the animal created by science would act in exactly the manner his creators train him to act.'

The creation of a Super Worker was probably the reason why the Soviets were willing to give Ivanov cash to pursue his research at a time when they didn't have a lot of cash to spare. In true revolutionary fashion, the communist leaders imagined artificially engineering a transformation not only in society, but in humanity itself. An ape–human hybrid was to be the bizarre but logical end point of their quest to produce a 'New Soviet Man'. As the hybrid man–apes toiled in the mines and fields, the humans would enjoy the carefree life of the Worker's Paradise.

After settling in at the Kindia station, Ivanov attempted to artificially inseminate several of the station's female chimps with human sperm acquired from an unnamed local donor. Thankfully, genetics worked against him. None of the chimps conceived. Ivanov returned to his homeland discouraged, hoping to find a nice Russian girl who would be willing to bear the child of Tarzan, the male orangutan he brought back with him. He actually found a few volunteers, but this time it was politics that worked against him. Suspected of harbouring counter-revolutionary sentiments, he was shipped off to a prison camp for five years. He died shortly after being released. There are rumours that Stalin ordered other Russian doctors to continue Ivanov's research, but there's no solid evidence to support such speculation.

The Stubborn Mentality of Apes

The French weren't the first colonial power to establish an ape colony in Africa. In 1913, under the auspices of the Prussian

Academy of Sciences, the Germans founded a primate research station on Tenerife, one of the Spanish-owned Canary Islands off the north-west coast of Africa. This station was far more important in the history of primate research than the French colony. However, because of the disruption of World War I, details about the work done there took a while to reach the outside world.

The Tenerife Station was conceived with lofty goals. Chimpanzees were to be taught to play musical instruments, speak German, and understand basic mathematics and geometry. Inevitably, hybridization also reared its head, with the German sexologist Hermann Rohleder expressing a desire to conduct human–ape breeding studies there. Many people outside Germany suspected the station had an additional, undisclosed mission. They figured it was an elaborate front for a German spy operation, the reasoning being that the location afforded an excellent view of transatlantic ocean traffic. According to this theory, when not giving the chimps violin lessons, the scientists would be busy radioing home details about the comings and goings of foreign ships.

It's possible the German government ordered the Tenerife researchers to do double duty as spies, though no smoking-gun document has ever been found to corroborate this. But the rest of the plan – music instruction, language lessons, and hybridization experiments – soon fell by the wayside. Control of the station fell under the humane directorship of twenty-six-year-old Wolfgang Kohler, who decided instead to undertake a study of the 'mentality of apes'. He was curious about questions such as: are chimpanzees capable of tool use? And are they capable of insightful learning? Many modern researchers trace the start of the science of primate psychology to Kohler's studies.

Kohler's experimental method was to present his chimp subjects with various 'food acquisition' puzzles to solve. Day after day, sweating under the hot tropical sun, he placed bananas or oranges just out of their reach – dangling from the ceiling, or lying temptingly a few yards outside their cage – and then observed, notebook in hand, as they tried to figure out how to secure the treat. Typically the solution involved some form of lateral thinking. For instance, the chimp might have to tug on a string to get the fruit, use a stick to manoeuvre it into his cage, or build a stack of boxes to reach it on the ceiling.

Kohler's results offered encouragement to the proponents of ape intelligence. He discovered that the chimps didn't rely merely on trial and error to solve the puzzles. Instead, they seemed to show flashes of genuine insight about the nature of the problems. Initially there was usually a period of frantic scurrying around when a chimp was first presented with a puzzle, but eventually the animal would pause and consider the situation. He might scratch his chin. Then would come the 'a-ha' moment. Kohler claimed you could almost tell the exact instant when it arrived. The chimp would step back. His eyes would light up, and he would proceed to solve the problem. Kohler believed that this was clearly human-type thinking.

Although Kohler found that the chimpanzees were highly intelligent and could solve all kinds of puzzles, it soon became evident to him they would be useless as workers. The main problem was their personalities. They were wilful and obstinate. If they didn't want to do something, nothing could make them do it. In particular, they refused to do anything that smacked remotely of work.

In his book *The Mentality of Apes*, published in English in 1927, Kohler offered an example of this stubborn streak. At the end of each day, after the chimps had been fed, fruit skins littered their enclosure. Usually it fell to the caretaker to clean up the mess. But one day Kohler figured, why not save the caretaker some work and make one of the chimps tidy up? So he showed his star chimp pupil, Sultan, what to do. Kohler described what happened next:

[Sultan] quickly grasped what was required of him, and did it – but only for two days. On the third day he had to be told every moment to go on; on the fourth, he had to be ordered from one banana skin to the next, and on the fifth and following days his limbs had to be moved for every movement, seizing, picking up, walking, holding the skins over the basket, letting them drop, and so on, because they stopped dead at whatever place they had come to, or to which they had been led. The animal behaved like a run-down clock, or like certain types of mentally-deficient persons, in whom similar things occur. It was impossible ever to restore the matter-of-fact ease with which the task had at first been accomplished.

The Tenerife Station closed in 1920, but the American researcher Robert Yerkes, who conducted similar experiments first in California and then in Florida, encountered an identical obstinacy. His chimp Julius would roll himself into a ball or start playing with his foot if asked to do something he didn't want to do. In fact, just about everyone who has ever worked with apes reaches the same conclusion – they are wild creatures that cannot be tamed. Passive obedience is not in their genes. As the primatologist Vernon Reynolds has put it, 'While a lot of civilized customs can be trained in, a lot of uncivilized instincts cannot be trained out. And the latter increase with the growing maturity of the ape.' This is why trying to make apes pick crops, work in mines, or serve as butlers is ultimately an exercise in futility. The apes are smart enough to know what to do, but they're also stubborn enough to refuse to do it.

The Rise of the Robots

After the 1920s, it became rare to encounter any serious suggestion of using apes as manual labourers. The idea just petered out, passing into the realm of fantasy and eventually re-emerging in science fiction. For instance, the fourth film in the *Planet of the Apes* franchise, *Conquest of the Planet of the Apes* (1972), explored a future in which humans train apes as butlers and street cleaners. (Eventually, of course, as fans of the films will know, the apes revolt and humans get their comeuppance.)

However, the disappearance of the ape worker can only partially be attributed to the recognition of their temperamental, untameable nature. Just as important was that people found an even better creature to fulfil their dreams of being waited upon by happy, obedient slaves. They found a creature willing to work tirelessly without complaint, glad to perform the most menial tasks, never raising a voice in complaint, never sulking or even slowing down. They found the robot.

The word 'robot', referring to a mechanical man, first appeared in a 1921 play, *R.U.R. (Rossum's Universal Robots)*, written by Czech playwright Karel Čapek. In Czech, the term *robota* means work or labour. The concept of artificial people spread quickly. By the end of the 1920s, companies were displaying the first crude robots to a fascinated public. In 1930, the Westinghouse Corporation proudly exhibited 'Rastus Robot', which it had designed to resemble an African-American field hand. As the *New York Times* reported, Rastus would 'sweep the floor, switch lights on and off, rise, sit down and talk' all at the touch of a button. Why bother with quarrelsome apes when such willing servants were available?

Of course, the problem with any slave – be it human, ape, or robot – is that the slave owner can never shake the nightmare that a day will come when his obedient workers rise up and revolt. As the possibility of robot servants has inched steadily closer to reality, robot rebellion has become an increasingly dominant theme in popular culture, portrayed vividly in blockbuster movies such as *The Matrix* and *The Terminator*.

When and if the machines do take over, don't expect our primate cousins to shed any tears for us. Given the history of how we've treated them, they'll probably be glad to see us go and will welcome the new robot overlords with open arms.

The Man Who Made Apes Fight

Guinea, West Africa – 1967. The troop of chimpanzees moves swiftly and silently through the tall grass of the savannah, travelling along a hard-packed path that weaves around jackalberry trees and acacia bushes.

They walk on all fours, often pausing to stand upright and scan what lies ahead. The troop includes ten females with young infants clinging to their fur. Several older infants scurry along under their own power. A large male, grey hair streaking his muscular back, leads the way.

They arrive at a clearing bordered by trees and pause. Then, looking to their left and right, they file out cautiously one by one. But suddenly there's a noise – a rustling of grass. They all look in the direction of the sound and freeze in shock. Somehow, despite their vigilance, they missed the danger directly in front of them. At the opposite side of the clearing, a large spotted leopard stretches out on the ground, basking in the hot afternoon sun. Between its front paws it holds a lifeless baby chimpanzee.

Their shock lasts only a second before the stillness of the savannah erupts into pandemonium as all the chimpanzees scream in unison and leap backwards. The next instant, they begin frantically rushing around, looking for weapons. Some of them grab sticks from the ground. Others break off branches from nearby trees and brandish them like clubs, six feet long, over their heads. Still screaming, and often pausing to pound the ground with their hands and feet, they form into a semicircle, 20 feet away from the leopard.

The leopard seems curiously unimpressed by the commotion. It moves its head from side to side, and swishes its tail back and forth, but otherwise remains where it is, as if reluctant to leave the meal it was about to enjoy.

The chimpanzees now begin to attack. Small groups of them, barking savagely, take turns rushing at the leopard. They approach to within several yards before retreating back to the safety of the group, repeatedly touching hands with each other for reassurance. One chimp rips a small tree out of the ground and swings it around and around like a whip. Others fling sticks and stones at the leopard. These missiles connect with their target with a painful smack.

Strangely, despite the barrage of sticks now piling up on its back, the leopard continues to lie still, making no noise. The only indication that it's even aware of the chimps is the constant movement of its head and tail.

Puzzled by the leopard's lack of response, the chimpanzees move closer. Their rushing attacks grow more frequent. The fusillade intensifies.

At last, a large male chimp creeps up to within several feet of the leopard. He watches carefully for any signs of movement, and then he darts forward, grabs the large cat by the tail, and charges back across the clearing, pulling the leopard behind him. It bounces along without resistance. After 30 feet, its head falls off, but the chimp continues on until he's dragged the body into the undergrowth.

The other chimps gather around the head, staring down at it in horror. There's no sign of blood. It actually still seems alive, staring up at them with wide, unblinking eyes. Barking ferociously, a chimp raises a large branch upwards and smashes it down on top of the atrocity. His companions join in the assault. Minute after minute they pound the leopard head, shrieking and barking, continuing to vent their rage upon it even after the eyes have fallen off, the seams have ripped apart, and there's nothing left but a pile of padding littering the ground.

During the 1930s, as a teenager in the Dutch city of Rotterdam, Adriaan Kortlandt would often get on his bicycle after school and pedal out to a nature park in the suburb of Lekkerkerk. He spent long hours alone there, watching and photographing a colony of cormorants, a pelican-like bird, as they built their nests and dove into the water to catch fish. He particularly enjoyed observing the young nestlings clumsily flapping their wings during their first flights, or learning how to hunt for food by searching for objects (wood, straw, string, paper) in the water. Birdwatching appealed to his solitary nature, because when he was around animals he felt at peace. People, on the contrary, often infuriated him.

After high school, Kortlandt attended Utrecht University, where he studied psychology and geology, but he continued his birdwatching. In the spring of 1939 he returned to the colony and built a tower, 12 metres high, which allowed him to sit near the birds unobserved. He was close enough to hear the soft sounds of the females as they mated. He spent five months alone in the tower, sleeping and eating up there, filling journals with cramped notes about the activities of the cormorants. He arranged to have his food winched over on a cable.

Kortlandt's passion for birdwatching, and his keen eye for detail, brought him to the attention of Nikolaas Tinbergen, a charismatic Dutch biologist, ornithologist, and pioneer of ethology, the scientific study of behaviour patterns in animals. The two struck up a correspondence, and Tinbergen tried to draw the moody but brilliant young man into the group of students and colleagues he was gathering around himself. At first Kortlandt responded to Tinbergen's overtures. He visited him several times, but soon his solitary nature prevailed. He found Tinbergen's research group to be too chummy. He complained it was like a football club. He also hated deferring to the older man's leadership. By this time, Kortlandt had decided to make a career in ethology, but he was determined to do it on his own terms.

Steadily the relationship between the two men deteriorated. In fact, it turned into a bitter rivalry – a clash of personalities that remains legendary among Dutch scientists to this day. Kortlandt and Tinbergen argued over everything, from the trivial (whether it was better to study seagulls or cormorants) to the profound (the proper degree of overlap between the studies of human psychology and animal behaviour). Truth be told, much of the rancour stemmed from Kortlandt's behaviour, since he provoked Tinbergen by making caustic remarks about him in print, as if purposefully challenging the authority of the senior scientist. Tinbergen responded by increasingly ignoring the brash young researcher.

The confrontation came to a head in 1941, when Tinbergen and one of his graduate students, Gerard Baerends, announced they had found an important organizing principle of animal behaviour. The instincts of animals, they argued, formed a pyramidal hierarchy, with a large base of lower-level instincts supporting a much smaller number of top-level instincts. When Kortlandt read this, he was overcome with anger. That was his theory! Or so he insisted. It was true that their conclusions were based on a study of digger wasps, not the birds that he studied, but he was adamant that he'd made similar observations several years earlier. Why hadn't Tinbergen and Baerends acknowledged his work?

Kortlandt complained to Tinbergen and Baerends, but he received no response. There turned out to be a good reason for this. The Nazis had shipped Tinbergen off to a prison camp because of his vocal support for Jewish colleagues. So Kortlandt temporarily let the matter drop, but he didn't forget about it. Instead, he nursed his grudge, and after the war he approached Tinbergen at a conference, demanding an answer. Tinbergen brushed him off, saying he was too tired to talk about it.

If Kortlandt could have let the issue go, his life might have turned out quite differently. Tinbergen was hugely popular in the Netherlands, hailed as a national hero because of his bold opposition to the Nazis. It wasn't smart to choose him as an enemy, but this didn't deter Kortlandt. He continued his complaints, and Tinbergen continued to ignore him.

The academic community rallied around Tinbergen. Kortlandt's confrontational manner made his scientific colleagues uncomfortable. They whispered behind his back when he attended conferences: *This guy has a chip on his shoulder! Watch out for him!* Kortlandt found himself increasingly marginalized within his profession because of his inability to play nice.

Then, one day in the mid-1950s, as Kortlandt wandered through the Amsterdam zoo, his gaze came to rest on a group of chimpanzees. Fascinated, he watched them milling around their enclosure. It seemed almost as if he were gazing through a time machine at the early ancestors of mankind. He pictured troops of chimpanzees wandering freely through the steamy solitude of the African jungle. It was a setting about as distant from the petty academic politics of the Netherlands as he could imagine, and Kortlandt knew right away that's where he wanted to be.

Shifting from cormorants to chimpanzees represented a major career change for him, but Kortlandt couldn't get the idea out of his head. He started reading up on primate behaviour. He inquired about the best locations to observe apes in the wild. And finally he made up his mind. He would leave behind the claustrophobic environment of the Netherlands and travel to the expansive jungles of Africa to study chimps!

Experiments with Chimpanzees in the Wild

In April 1960, at the age of forty-two, Kortlandt arrived in the eastern Congo, ready to start the new phase of his career. The project he was embarking on was groundbreaking. Richard Garner, the 'simian linguist', had attempted to study apes in the wild at the end of the nineteenth century, but his methods were considered peculiar and unscientific. Ever since then, no one had conducted a systematic field study of apes. Part of the problem was that it cost a lot of money to travel to Africa. If a researcher wanted to study chimpanzees or gorillas, it was far cheaper to observe captive specimens in a laboratory or zoo. So Kortlandt's Congo expedition, made possible by a decline in air travel prices, was a bold new initiative. In fact, it represented the beginning of modern field primatology.

Kortlandt based himself at a banana and papaya plantation, where the local wild chimpanzees often wandered out of the forest to feed. He built a camouflaged hunter's hide in which he could spy on the apes without their knowledge. He also placed an aerial observation post 80 feet high in a tree, which afforded him a panoramic view of the entire terrain. Then he waited for the chimps to show up.

He usually had to wait for hours, and he would hear them before he saw them – the cries of the adult males rising up from the jungle. Finally, a face would appear, peering out from the foliage at the edge of the clearing. Black eyes would cautiously look left and right, making sure all was safe. Then a chimp would step out, as if materializing out of the leaves. One by one, more followed. When the entire troop had regrouped, they would happily run and play, chasing each other around, stomping the ground with their feet, and pulling fruit down from the trees. Kortlandt crouched alone in his hunter's hide, taking notes on everything he saw.

The hidden observation wasn't without danger. Kortlandt related how one time, while clinging to his aerial observation post, a storm came in and began violently blowing him back and forth. Lightning flashed on the horizon. Despite the risk of electrocution, Kortlandt stayed where he was, unwilling to reveal his presence to the apes. He figured that if he died, at least he would die without ruining his research.

Not content to observe the chimpanzees passively, Kortlandt soon decided to turn the forest into a massive laboratory in which he would conduct tests. He described these as his 'experiments with chimpanzees in the wild'. An academic reviewer of his work later referred to them as his 'zany field experiments'.

Kortlandt began by surreptitiously placing test objects around the forest and recording the reactions of the chimps who stumbled upon them. For instance, in order to find out about chimpanzee eating habits, he set out samples of food such as hen's eggs, mango, grapefruit, and pineapple. He observed that the chimps carefully avoided food unfamiliar to them. From this he concluded they were picky eaters who based their dietary choices on what they knew, rather than on instinct or smell.

A more gruesome experiment involved placing the bodies of dead, unconscious, or dummy animals around the jungle. His inspiration for this was a 1946 study conducted by Donald Hebb at the Yerkes Anthropoid Center in Florida. Hebb had discovered that the sight of 'inert, mutilated or dismembered bodies' elicited a strong, and seemingly innate, fear reaction from chimpanzees. Even a human or chimpanzee head modelled out of clay sent them into a panic. These reactions suggested to Hebb that, unlike any other species except humans, chimpanzees understood the difference between life and death.

To find out if he could replicate Hebb's results, Kortlandt set out a series of objects in the jungle that included a dummy squirrel, a stuffed mouse, a dead bird, an anaesthetized black goat, and a similarly drugged mangabey monkey. The apparently dead mammals definitely scared the chimps – more so, in fact, than real threats such as giant spiders – but the object that terrified them the most was a lifelike portrait of a chimpanzee's head mounted on a piece of cardboard. The chimps stared at it in horror from a distance, as if observing some kind of hideous apparition. Finally, they made a wide detour around it.

Armed Fighting Techniques of Wild Chimpanzees

As Kortlandt watched the chimps, he was constantly amazed at how human they seemed. But the question that really intrigued him was: why weren't they *more* like humans? Why, when the human and chimpanzee lineages diverged 5 to 8 million years ago, hadn't chimpanzees continued along an evolutionary path similar to that of humans? Why had our ancestors gone on to build cities and empires, while their ancestors continued to look for fruit in the jungle? Eventually Kortlandt came up with an answer. He called it the 'dehumanization hypothesis'.

Kortlandt's theory was based on geography. He noted that humans evolved on the open savannah. This environment favoured walking on two legs to see above the grass, as well as the development of aggressive hunting behaviour and the use of weapons because of the lack of readily available fruit. Chimpanzees, however, evolved in the jungle, where a vegetarian way of life made more sense, and where the use of weapons was impractical because it was difficult to swing a club or throw a spear without hitting a tree.

Kortlandt imagined a time, millions of years ago, when the ancestors of modern chimps and humans might have lived together on the savannah. It was at that time, he argued, that the proto-chimps acquired many of the human-like characteristics still evident in their modern descendants. However – and this was the crucial part of his hypothesis – when the early humans learned to use spears, they drove the chimpanzees back into the jungle.

Trapped there, chimpanzees lost the skills they had acquired on the savannah. 'In the forest,' Kortlandt wrote, 'the semihuman element in the behaviour of apes faded away to a large extent.' Chimpanzees became 'dehumanized'.

It was an intriguing idea, but Kortlandt needed a way to test it. He reasoned that if he was correct, then forest and savannah chimpanzees should fight in different ways. Both should instinctively use weapons, but savannah chimps, having relearned the skills of their ancient ancestors, should use them far more effectively than their forest-living counterparts. The problem was, how could he observe the fighting behaviour of chimpanzees? Somehow he would have to provoke them to attack. He puzzled over how to do this for a long time, and then, in a flash of inspiration, the solution came to him. He'd use a stuffed leopard!

The experiment Kortlandt imagined was to confront the chimpanzees with a life-size, animatronic model of a leopard holding a baby chimpanzee doll in its claws. Since leopards were the chimps' main enemy, the sight of one, he hoped, would trigger their combat instincts. The concept was creative, but it also was a curiously fitting expression of Kortlandt's own personality. Having left the academic disputes and quarrels of the Netherlands far behind, he was now scheming to provoke fights with the new group of primates he found himself among.

Kortlandt obtained a large stuffed-toy leopard as a donation from the Panther Cigar Company and shipped it out to the Congo. He then placed the leopard at the edge of a clearing, along a path the chimps often took to search for food. The model was on a sled, allowing it to be rapidly pulled out of the bushes when the chimps approached. A windshield-wiper motor installed inside its body moved its head and tail, making it seem more alive. It resembled something you'd encounter on a retro jungle ride at Disneyland, but it was lifelike enough to fool the chimps. Kortlandt hid in a tree house 40 feet above the clearing, operating the controls of the leopard. With him was a cameraman to catch the entire scene on film.

The two men waited. Finally a troop of chimpanzees wandered out into the clearing. Kortlandt pulled on the rope, and the leopard model slid out from its hiding place. He described what happened next:

> There was a moment of dead silence first. Then hell broke loose. There was an uproar of yelling and barking, and most of the apes came forward and began to charge at the leopard. . . . Some charges were made bare-handed, in others the assailants broke off a small tree while they ran towards or past the leopard, or brandished a big stick or a broken tree in their charge, or threw such a primitive weapon in the general direction of the enemy. They beat the ground and stamped on it with hands and feet, they struck out and kicked at tree trunks, they used flexible trees, still rooted, as a whip by lashing out with them at the beast of prey, often from a very short distance, and they climbed up into trees, and shook them so violently that once a papaw tree of 6 metres broke and, together with the ape in it, fell to the ground.

Kortlandt later boasted that the display was 'the greatest show on earth ever seen in the field of animal behaviour'. It took the chimps about half an hour of running around and screaming before they figured out the leopard wasn't real. Once they did, they poked at it cautiously and smelled it. One of them sat down in front of it, dismissively turned his back on the strange creature and ate a banana. Eventually the chimps lost interest in the non-responsive leopard and drifted back into the forest. They returned once more to make sure the creature was still dead, and then made a final departure.

Despite all the screaming and drama, the important point was how the chimps used their weapons. Kortlandt observed that they definitely brandished sticks and branches, but as he had predicted, they were extremely clumsy with them. Not one of the branches they hurled succeeded in hitting the leopard. Kortlandt conducted five more trials with the forest chimps and always obtained similar results. The next step was to repeat the experiment with savannah chimps.

These proved much harder to find, since the people who lived in savannah regions tended to drive the apes back into the jungle (again, as the dehumanization hypothesis predicted). But eventually, in 1967, Kortlandt found a group of wild savannah chimps living in a region of Guinea where the local population's religion forbid hunting them. He organized an expedition to restage the experiment.

This time, the difference was dramatic. Like the forest chimps, as soon as the savannah apes saw the faux leopard they picked up branches and brandished them as weapons, but what they did next demonstrated far greater skill and organization. They formed a semicircle around the leopard, and, one by one, took turns charging it and pelting it with sticks – with deadly accuracy. By later calculations based on the film of the event, Kortlandt estimated they threw the missiles at speeds over 60 mph. Finally one chimp rushed forward, grabbed the leopard by the tail, and dragged it into the undergrowth, causing its head to fall off. The chimps stared in horror at the bodiless head, and then started pummelling it savagely. They only calmed down once the sun began to set, at which point they slunk away into the darkness.

Kortlandt felt the experiment had decisively confirmed his dehumanization hypothesis. The forest and savannah chimps clearly displayed different levels of ability with the weapons. In November of that year, he triumphantly showed footage of the tests at a conference in Stockholm. The conference attendees sat speechless as they watched the enraged apes smashing apart the stuffed leopard. No one had ever seen anything quite like it before. The footage made headlines around the world.

Unfortunately for Kortlandt, acceptance of his hypothesis didn't follow. Rival scholars agreed his theory was interesting, and that the footage was dramatic, but they questioned whether the human-like characteristics of chimpanzees couldn't be explained in other ways.

Surely such traits could have evolved in the jungle. And what did a stuffed leopard really prove? Would chimpanzees have attacked a live leopard in the same way?

A Leopard in a Ball

Part of the reason why the dehumanization hypothesis didn't attract a wider following was, of course, Kortlandt himself. His social skills hadn't improved any during his years in the jungle, and his aggressive manner, which was often interpreted as arrogance, undermined his attempts to make allies in the scientific community. As his colleague Jan van Hooff tactfully put it, 'Kortlandt has always been an individualist with a markedly critical and provocative stance that has sometimes hindered the acceptance of his ideas and their merits.'

Kortlandt didn't let the chilly response discourage him. Instead, he treated it as a challenge. If a fight between a live leopard and a troop of chimps was the only thing that would convince the scientific community, then that's what he'd give them, and a few years later he figured out a way to achieve this.

In 1974, Kortlandt transported a tame leopard from the Netherlands to Guinea. Once there, he placed it inside what he called a 'leopard launching construction', which was a large wire-mesh ball he had trained the leopard to 'roll-walk' inside of. His plan was to launch the leopard from the top of a hill. It would then roll down to the chimpanzees who, if all went to plan, would attack it. Kortlandt hoped that the wire-mesh would protect the leopard, because otherwise it was going to be defenceless.

It would have been a fascinating sight to see, and would have produced thrilling footage, but unfortunately a devastating brush fire tore through his camp, destroying much of his equipment and ending the experiment before it took place. So the world never found out how savannah chimps would respond to the sight of a live leopard rolling down a hill towards them in a wire-mesh ball.

The setback was doubtless upsetting for Kortlandt. Adding insult to injury was the news that his old rival Tinbergen had been awarded the Nobel Prize the previous year. The award was given, in large part, to honour theories Kortlandt continued to believe were rightfully his own.

A Solitary, Thorn-Swinging Sentinel

Although Kortlandt continued to feel he wasn't receiving the acknowledgement he deserved from the scientific community, he soldiered on, dreaming up new field experiments to prove his theories. For instance, his critics questioned how early hominids could have survived on the savannah against predators such as lions. At first, this stumped Kortlandt. Early hominids, before they learned to use spears, surely would have been easy prey for lions. The answer came to him while hiking through Kenya's grasslands. He noticed thorny acacia bushes growing everywhere. Snapping off a spiky branch, he imagined a protohominid using it as a defensive weapon, swinging it around like a whip to drive away predators.

Clearly, this called for another experiment. Initially he envisioned donning a suit of metal, using himself as bait, and driving off the lions with thorns. But he soon decided that the sight of an armoured scientist stalking the savannah would scare away the lions. Instead he turned to his favourite piece of technical apparatus, the windshield-wiper motor, and built a device like a miniature helicopter rotor, with thorn branches attached to the blades. From the safety of his car, he could press a button and cause the branches to spin around, smacking any approaching predator on the face.

In October 1978, on the estate of George Adamson, of *Born Free* fame, he erected his thorn-swinger above a tempting plate of raw steak. Soon the lions arrived and warily circled the meat. For almost an hour the device stood guard over the food – a solitary sentinel, whipping its thorny branches at any who dared approach. If Kortlandt had been trying purposefully to create a mechanical self-portrait, he couldn't have done a better job. Finally one of the lions gathered its courage, crept up on its stomach, and reached out for the meat. *WHACK!* A thorn branch smacked it right on the nose. The lion leapt back in a panic, and the entire pride scattered into the undergrowth.

Again, Kortlandt proudly showed footage of the experiment to his colleagues. Jan van Hooff commented that the bizarre film stuck on the retina of everyone who saw it. But following the by now customary pattern, there was a certain reluctance among the scientific community to take Kortlandt seriously. His experiments were just a little too weird, and his personal manner just a little too abrasive.

The Corpse in the Closet of Behavioural Biology

It's interesting to contrast Kortlandt's career with that of Jane Goodall, who conducted similar research at around the same time, but quickly earned far more acclaim and respect from the scientific community. Just three months after Kortlandt began his work in the Congo in 1960, Goodall arrived in Tanzania, where she launched a study of the wild chimpanzees of Gombe National Park. But although Goodall and Kortlandt shared the goal of studying wild chimpanzees, they went about it in radically different ways. Kortlandt hid from the chimpanzees, terrified them with frightening objects, and lured them into armed struggle. Goodall, by contrast, tried to gain the trust and acceptance of the animals. Instead of hiding from the apes, she openly approached them. First she sat at a distance, passively observing them, habituating them to her presence. Gradually, day after day, she inched closer. Weeks later, she was right beside them. This allowed her to obtain naturalistic observations of primate behaviour at close hand.

Given a choice between idyllic scenes of an attractive young woman peacefully interacting with apes, and Kortlandt's surreal scenes of leopard-fighting chimps, it's small surprise that the public preferred the former. Goodall became an international scientific celebrity, rocketed to fame by a series of *National Geographic* documentaries about her work. Kortlandt remained obscure and bristled at any comparison to Goodall, complaining to anyone who would listen that her work lacked scientific rigour.

For the remainder of his career, Kortlandt let his grievances fester. Finally, in 2004, at the age of eighty-six, he convened a press conference to vent his anger, luring reporters to his small Amsterdam apartment with the promise that he would reveal the 'denouement of a scientific mystery'. The mystery, it turned out, was why he'd never received the fame and awards he thought were rightfully his.

Two reasons prompted him to seek out the press. First, he had offered to speak at the European Conference on Behavioural Biology, held that year in Groningen, but the conference leaders turned down his offer, evidently fearing he'd use the opportunity to rant about Tinbergen, who by that time was dead and couldn't defend himself. Kortlandt wanted to protest the injustice of this rejection. He insisted he had merely wished to give a farewell address to the scientific community. Second, he'd completed the first volume of his autobiography, a projected three-volume series, but had been unable to find a publisher. Again, he felt, this was more evidence of a conspiracy against him.

The journalists sat patiently in his living room, listening to him recite his laundry list of affronts and insults. A reporter for *De Volkskrant* newspaper wrote that a storm was raging outside, but that the crashing of the thunder couldn't match the fury of the old man inside. 'I am the corpse in the closet of behavioural biology,' Kortlandt cried out at one point.

Kortlandt passed away five years later, at the age of ninety-one. He had continued to work on his memoir up until the very end, confident it would finally set the record straight, perhaps imagining each line landing like a thorny acacia branch, inflicting stinging, painful revenge on all his enemies.

The Mystery of the Missing Monkey Orgasm

The Yale Laboratories of Primate Biology, Orange Park, Florida –
June 1937. A wire-mesh partition lifts with a metallic clang, opening a
passage between two adjoining cages. The occupant of one of the cages, a
large male chimpanzee named Bokar, pokes his head inquisitively through
the new door. His eyes connect with the occupant of the other cage, a
mature female chimp named Wendy who's clinging to the netting
attached to the wall. Bokar hoots and slaps the ground with excitement.
Wendy responds with a low grunt, and then glances away coyly.

Bokar gestures to Wendy with his hand: COME HERE! Wendy looks
intrigued, but doesn't move, so Bokar gestures again, touching his genitals to make his intentions clear. This time, Wendy responds positively.
She climbs down from the netting, walks towards Bokar, and lies down
on the floor, offering herself submissively to him. Bokar approaches and
sniffs appreciatively.

Outside the cage, two researchers, a man and a woman both dressed
in white lab coats, observe the behaviour of the chimpanzees. The man
speaks in a hushed voice to the woman.

'We record everything that occurs on this standardized scoring sheet
of mating behaviour. For instance, Wendy just lay down on the floor.
That's considered "low presentation". It's a sign of sexual receptivity. So
we check the box here.'

The woman nods to show that she understands, and the male scientist continues. 'Ah, Bokar is displaying an erection. We check here. And
copulation is about to begin. Chimpanzee mating, you'll find, is quite a
rapid affair. It often lasts no more than two or three minutes. Now we
need to count the number of thrusts.'

A few seconds pass. 'How many did you count?'

'Fourteen,' the woman replies.

'Good. So did I. And ejaculation?'

'Yes.'

'We check here.'

'What about the female?' the woman asks.

The man looks at her quizzically. 'What do you mean?'

The woman blushes slightly. 'Should I record if the female has a climax?'

For a moment the man looks confused. Then he snorts derisively. 'That's an odd question. To be honest, I've never seen such a thing, but it's certainly not relevant to our studies.'

The woman glances down at the floor. 'I'm sorry. I didn't mean to question the protocol.'

'Not a problem. Just follow what's on the scoring sheet and you'll do fine.'

The man returns his attention to the chimpanzees. 'Ah, I see Bokar has begun again. Now watch closely and please count his thrusts.'

Early in the twentieth century, the psychologist John Watson noted with dismay the difficulties facing anyone foolhardy enough to study human sexuality. Though the topic was widely admitted to be of great importance, any intrepid researcher who chose to tackle it had to deal with social taboos, personal attacks, and human subjects who might be either overly modest or outright deceptive. As a result, most scientists steered clear of the topic.

One solution to this problem was simply for researchers to develop a thicker skin, and to proceed forward in spite of the shocked cries of critics. The pioneer sexologists – Watson, Alfred Kinsey, Virginia Johnson, and William Masters – all adopted this approach.

However, there was a second possible solution, one more subtle and indirect. This was to study the sexual behaviour of primates, and to extrapolate from them to us. Advocates of this method noted it offered a variety of advantages over the direct study of humans. First, primates were forthright creatures. Unlike people, they weren't going to lie about their sex lives. Second, apes and monkeys seemed free of the cultural baggage of humans. They were like simpler versions of ourselves – the way our distant ancestors, the protohominids, might have been before civilization added layers of repression and neurosis on top of their instinctual urges. By studying primates, it might be possible, therefore, to peer deep into the primitive roots of human behaviours. Lastly, researchers could conduct experiments with primates that, for ethical reasons, weren't possible with humans. So it was, during the first half of the twentieth century, that a handful of researchers turned their curious eyes to the copulation of primates.

At first, everything went according to plan. The correlation between primates and humans seemed relatively straightforward, but then, as so often happens in scientific research, a complication emerged.

Thanks to biology, male primate desire was easy to measure and record. The feminine response, however, proved more inscrutable. Specifically, how was a human researcher supposed to know if a female primate was having an orgasm? It's hard enough for a human male to know that about a human female, so imagine the difficulty with primates. Researchers couldn't ask them, and what external signs could they look for? Eventually the suspicion was verbalized: *Do female primates even have orgasms?* Thus arose one of primatology's more curious controversies – the debate over the missing monkey orgasm.

Monkeys in Montecito

The study of primate sexuality began in 1908, when the psychologist Gilbert Van Tassel Hamilton moved from chilly Boston to sunny Montecito, located just outside of Santa Barbara in Southern California. He had secured a comfortable position as the personal physician of Stanley McCormick, heir to the fortune of Cyrus McCormick (founder of the McCormick Harvesting Machine Company). Stanley had some rather florid psychological problems that included manic depression, violent rages, impotence in the presence of his wife, and compulsive public masturbation. Stanley had even resorted to wearing a leather crotch harness in an attempt to prevent his self-pleasuring urges.

In the idyllic setting of Montecito, surrounded by the oak trees at the edge of the Los Padres forest, Stanley's symptoms improved. This allowed Hamilton the free time to conduct his own research. Though he had agreed to be McCormick's personal physician, he didn't want to abandon his academic career entirely. He was particularly eager to pursue his interest in the psychology of abnormal sexual behaviour. The problem was how to go about doing that in the Southern California countryside, with access to only one patient.

It occurred to Hamilton that monkeys might be the answer to his problem. The temperate climate of Southern California would make an ideal location for a monkey colony, and by studying their sexual tendencies, he might gain valuable insights into the biological origins of abnormal human behaviours.

Hamilton discussed his ideas with the McCormicks. He probably pointed out to them that such research might shed light on Stanley's condition, and soon he had persuaded them to import a colony of twenty-six macaques and two baboons for his use. The animals were given a large yard in which they could run, play, and copulate. And so Hamilton became a trailblazer of the study of primate sexuality, a subject about which almost nothing had been known previously.

Hamilton was gratified to find that the monkeys offered copious examples of abnormal behaviours. For instance, Hamilton observed a female named Maud who became sexually aroused, smacking her lips excitedly, whenever a dog entered the yard. Unfortunately for Maud, the dog interpreted her advances as a threat and bit her arm. After that, all the monkeys carefully avoided the dog.

Intrigued by these displays of interspecies sexual desire, Hamilton designed an experiment to test how far such tendencies would extend. That is, what variety of creatures would a monkey voluntarily try to have sex with? He placed Jocko, a large male monkey, alone in a cage, and then supplied him with a series of different animals – first a gopher snake, then a kitten, a puppy, and finally a fox. Hamilton happily reported, 'He attempted sexual relations with each of them after he had inspected their genitalia.'

For Stanley's sake, Hamilton was particularly keen to observe the masturbatory habits of his monkeys, but to his surprise, they hardly ever displayed that behaviour. Only Jocko (the animal lover) engaged in that pastime, and then only when he was confined for long periods. Hamilton began to suspect Jocko might be a bit more abnormal than the typical monkey, but in any case, this left Hamilton at a loss to explain the biological roots of Stanley's condition. Poor Stanley seemed to be some kind of evolutionary outlier.

As for whether the female monkeys were having orgasms, Hamilton never mentioned the subject. The question probably never occurred to him.

Hamilton's work with the monkey colony ended abruptly in 1917, after Stanley suddenly relapsed. The McCormick family blamed Hamilton for the setback. Perhaps they had also grown tired of living with a bunch of libidinous monkeys. Whatever the reason may have been, they fired him. So Hamilton, doubtless feeling dejected at the loss of his colony, moved back to the east coast and joined the military.

Experimental Matings at the Monkey Farm

As a result of Hamilton's falling-out with the McCormick family, primate sex research was disrupted for a number of years, but eventually it continued and expanded thanks to a former colleague of Hamilton's from Boston, the psychologist Robert Yerkes (whom we met, briefly, earlier as he was engaged in a clash of wills with his stubborn chimp Julius).

In 1915, Yerkes had spent half a year with Hamilton at Montecito. The experience inspired him to try to create a full-scale primate laboratory in the United States. Throughout the 1920s, he sought ways to make this happen. Finally, in 1930, with financial help from the Rockefeller Foundation, Yerkes founded the Anthropoid Experiment Station of Yale University, in Orange Park, Florida.

Local residents referred to it as the Monkey Farm. At its opening, it housed seventeen chimpanzees and sixteen other apes. The facility, now officially called the Yerkes National Primate Research Center, still exists, though it has been relocated to Georgia, and remains one of the leading primate labs in the world.

Unlike Hamilton, Yerkes wasn't particularly interested in abnormal behaviours. Instead, he wanted to identify precisely what was normal. It was part of his grand project for the scientific management of society. He planned to use laboratory chimps as a stand-in for humans, gleaning knowledge from them that he would then apply to the rational organization of mankind. Sex research was only a part of this project – from chimps, he promised, would flow broad improvements in education, medicine, and psychology – but it was an important part. He envisioned that intensive study of the reproductive life of apes would yield 'fruitful application in human social biology'.

To fulfil this plan, Yerkes and his staff conducted experimental matings in which they caged male and female chimps together, hoping to pinpoint the perfect mixture of male desire, female receptivity, and mate acceptability that would create ideal reproductive conditions. Between May and August 1937, 233 of these matings took place. Clipboards in hand, the researchers carefully observed the different chimp couples and measured their sexual interactions. In an effort to make the study as objective as possible, Yerkes distilled the act of ape sex down to a series of items on a standardized scoring sheet that were checked off as the anthropoid excitement mounted.

> Male approaches – check.
> Grooming – check.
> Female presentation – check.
> Genital examination – check.
> Erection – check.
> Number of thrusts – 21.
> Ejaculation – check!

It was from these experimental matings that the lacuna of the female orgasm eventually arose, ghost-like, into scientific view. However, it wasn't Yerkes who first noticed its absence. Ruth Herschberger, a New York psychologist and early feminist, wrote to Yerkes in 1944 objecting to a persistent male bias she perceived in his work. For instance, she pointed out that he described male apes as being 'naturally more dominant' except when females went into oestrus, at which point the females appeared to become more dominant. Did this, Herschberger asked, mean that he thought female dominance was unnatural? Yerkes grumbled that his terminology had been 'convenient'.

But Herschberger continued. Why did Yerkes assume that female apes were such sexually passive creatures? That is, why did he measure male desire, but merely female 'receptivity'? Why not measure female desire as well? And for that matter, where did the female orgasm appear on his checklist? It wasn't there. Why not? Herschberger pressed him on the issue. Had Yerkes or any of his researchers, she asked, ever observed a female chimp having an orgasm?

Yerkes admitted Herschberger posed an interesting question, but he didn't have an answer. Despite the hundreds of matings his researchers had witnessed, he couldn't say for sure whether they had ever seen a female orgasm – and, of course, it wasn't possible to ask the chimps.

The Wrong Position

Once the question of the missing female primate orgasm had been raised, it couldn't be ignored. Where was it? Did it exist? At first blush, these may have seemed like ridiculous questions. After all, why wouldn't female primates have orgasms just like their human counterparts? Female primates have a clitoris, an organ with no apparent purpose other than sexual stimulation. So where there's a clitoris, isn't it safe to assume there's also an orgasm? Primatologists, however, weren't ready to make so bold a leap. Science moves slowly and carefully. It hesitates to jump to conclusions, especially with topics as explosive as female orgasms. And the fact remained that there was no solid, first-hand evidence of a female orgasm.

There was another reason for caution. From an evolutionary point of view, the female orgasm had always been a bit of a puzzle. What purpose did it serve? Unlike the male orgasm, it wasn't necessary for reproduction. So perhaps the female orgasm had never developed among apes and monkeys. The possibility couldn't be ruled out.

In 1951, Yale psychologists Frank Beach and Clellan Ford authored *Patterns of Sexual Behavior*, an encyclopedic study of sexual behaviour throughout the animal kingdom. In it, for the first time in the scientific literature, they directly addressed the question of the female primate orgasm.

Beach and Ford first summarized what was known about the primate sex act: 'The receptive female primate appears to indulge in one coital act after another, giving little or no indication that she experiences any explosive, temporarily exhausting, release of built-up excitement.' They then reviewed primate research to determine if anyone had ever seen anything resembling a female orgasm. They found only one slender piece of evidence. Clarence Ray Carpenter, while observing free-ranging monkeys on Cayo Santiago in 1940, had noticed that near the conclusion of the mating act, female rhesus macaques typically reached back and gripped their partner. He called it a clutching reflex. Perhaps here was evidence of the elusive orgasm.

Beach and Ford, however, weren't convinced. They acknowledged that female primates seemed to have the biological equipment necessary for an orgasm, but the mechanics of anthropoid copulation suggested that the clitoris, if it worked, wasn't being stimulated. This was because primates habitually engage in coitus by means of rear entry, and sex researchers in the 1950s believed that this position was insufficient to produce a female orgasm. The two researchers noted, 'Clitoral stimulation is minimal or lacking when coitus is accomplished by rear entry.' Therefore, they concluded, 'Their method of mating deprives them of one very important source of erotic stimulation.'

Beach and Ford delivered their final verdict: 'Orgasm is rare or absent in female apes and monkeys.' The sentence was heavy with the weight of scientific authority. Who could possibly argue with it? The two researchers had looked for the female orgasm among primates, given it every chance to show itself, and only then had dismissed it, as if it was the biological equivalent of a paranormal event. The phenomenon couldn't be seen, therefore it didn't exist. Rumours of its existence weren't enough to justify official acknowledgement.

A Female Perspective

For nearly twenty years, Beach and Ford's conclusion remained unchallenged. But then, like a repressed memory, the question of the monkey orgasm again rose to the surface. It wasn't any change in primate behaviour that prompted its return. Instead, it was changes in human society.

By the 1960s, women were entering the workforce in increasingly large numbers. This was true for science as well as other professions, and it was particularly true for primatology. Whereas other disciplines, such as physics, remained defiant preserves of masculinity, women seemed drawn to primatology. By the early 1970s, the two most famous primatologists in the world were women – Dian Fossey, renowned for her work among the mountain gorillas of Rwanda, and Jane Goodall, equally celebrated for her studies of the chimpanzees of Tanzania's Gombe National Park.

As the new female primatologists surveyed the previous half-century of studies, they couldn't help but notice, like Herschberger before them, a persistent male bias. For instance, male primatologists had designated Man the Hunter as the driving force of human evolution. But why was hunting so much more important than the gathering and caregiving work of early hominid women? Also, male researchers such as Yerkes had argued that male dominance was the binding force in primate society. This seemed conveniently to imply that male dominance in human society was the natural order of things and shouldn't be questioned.

Then there was the issue of the female primate orgasm. Why had male researchers been so quick to deny its existence? Perhaps the truth was not that female primates didn't have orgasms, but rather that male researchers hadn't bothered to look closely enough for them. Female researchers resolved to revisit the issue.

A 1968 study by Doris Zumpe, in partnership with Richard Michael, put the first dent in the case of the non-orgasmic female. Following up on Carpenter's observation of the 'clutching reflex', the two researchers patiently observed and filmed 389 rhesus monkey copulations. Then they painstakingly conducted a frame-by-frame analysis of the copulations.

Just as Carpenter had described, the females reached back and clutched the males when they became excited. A picture accompanying Zumpe and Michael's article shows a female monkey firmly holding her partner's snout in her paw while he enters her from behind. But what Zumpe and Michael found to be most significant was that the females did this *before* the males ejaculated. If the female failed to clutch the male, often he was unable to continue. Why would female clutching trigger male orgasm? Zumpe and Michael suggested that the female instinctively reached back when she was having an orgasm, and if this was the case, perhaps she was simultaneously experiencing vaginal contractions. Such contractions could give the male the stimulation necessary to finish.

It was a plausible argument, but even Zumpe and Michael acknowledged their evidence was more suggestive than conclusive. They were merely speculating about the existence of vaginal contractions. More specific measurements of physiological changes in the females were needed, but gathering these measurements would necessarily be intrusive and interfere with the natural mating process. After all, monkeys, unlike some humans, don't like to have sex when they're wired up to a machine. It wasn't clear how to avoid this problem.

Sexual Climax in Macaca mulatta

Although primate sex research had originally been intended to shed light on human sexuality, the opposite turned out to be the case. All those researchers who had chosen to develop a thick skin and directly tackle the study of human sexuality, without shame or embarrassment, had by the late 1960s gathered volumes of data on the physiological characteristics of the human female orgasm. In particular, the researchers Virginia Johnson and William Masters had determined, with the assistance of 700 male and female subjects, that the typical human sexual response cycle consisted of four phases: excitement (initial arousal), plateau (heart rate quickens, pleasure intensifies), orgasm, and resolution (the body relaxes and further stimulation may become painful).

In 1970, it occurred to anthropologist Frances Burton of the University of Toronto that she could apply Masters and Johnson's human sexual response data to the question of the female primate orgasm. Specifically, if she could demonstrate that a female monkey exhibited all the stages of the sexual response cycle, this would strongly imply that the monkey had experienced an orgasm.

The method Burton devised to achieve this proved to be ethically controversial. She didn't physically harm any monkeys. Nevertheless, many critics found the experiment to be disturbing from a psychological perspective. Burton strapped female monkeys in a restraining harness and used a plastic faux monkey penis to simulate copulation while monitoring the animals' physiological responses. As the historian Donna Haraway has noted, this experimental design verged on a 'caricature of rape'. The added irony was that Burton conducted the research under the banner of bringing a female perspective to primatology.

Burton tried to follow an exact protocol. The plan was to groom the monkeys for three minutes to put them at ease, and next to administer precisely timed periods of stimulation separated by intervals of rest. The monkeys didn't entirely cooperate. One of them grew so aggressive that she had to remove it from the study.

But somehow Burton managed to get two other monkeys to relax long enough to observe them progressing through phases of sexual response similar to those described by Masters and Johnson. Orgasm appeared to have been achieved.

It was hard to dispute Burton's results. Burton noted that the only possible way she could have enhanced her experimental design would have been to insert telemetric implants into the brains of her subjects in order to record their brain activity while she stimulated them. She offered this as a possibility for future research.

However, while Burton had settled the question of whether female monkeys *can* have orgasms, she hadn't addressed, as she herself acknowledged, the question of whether female monkeys typically *do* experience orgasms under natural conditions. Burton had stimulated her monkeys for over five minutes before getting the desired result, but monkey copulation in the wild usually lasts only a fraction of this time. To address this issue, the wheels of scientific research kept on turning.

Yes, They Climax, but Why?

Several years after Burton's study, Suzanne Chevalier-Skolnikoff spent five hundred hours at Stanford University's primate lab observing stumptail monkeys – a species unusual because of its 'striking orgasmic pattern'. When a male stumptail has an orgasm, his entire body goes rigid. This is followed by spasms. Then he makes a round-mouthed facial expression – imagine a look that combines awestruck wonder with complete befuddlement – while grunting loudly. During her observations, Chevalier-Skolnikoff witnessed numerous instances of female stumptails engaged in homosexual activity. During these trysts, the females displayed behavioural reactions similar to those of the males: rigidity, lip-smacking, and the round-mouthed expression. Based on these behaviours, it seemed reasonable to conclude the females were experiencing orgasms.

Throughout the 1970s and 1980s, other reports trickled in of orgasmic behaviour in female primates observed under naturalistic conditions. David Goldfoot implanted a battery-powered radio transmitter into a female stumptail macaque to remotely monitor her uterine contractions. He found the strongest contractions coincided with the moment during sexual episodes when she was making the round-mouthed face of wonder and joy, which was strong evidence of orgasm. Jane Goodall also observed female chimpanzees in Gombe masturbating, 'laughing softly as they do'.

Based on this evidence, when Alan Dixson sat down in the late 1990s to write the textbook *Primate Sexuality* he arrived at a conclusion with respect to female primate orgasm that was almost the exact opposite of the one Ford and Beach had reached half a century before. Dixson wrote:

> Although female primates exhibit behavioural responses indicative of orgasm less frequently than males . . . there is now plenty of evidence that physiological responses similar to those which occur during orgasm in women also occur in some macaques and chimpanzees. . . . Orgasm should therefore be viewed as a phylogenetically ancient phenomenon among anthropoid primates; the capacity to exhibit orgasm in the human female being an inheritance from ape-like ancestors.

So it took a while, but science finally did accept female primates as full-fledged members of the orgasm club. But as often happens in science, when one question is answered, another simply appears in its place. In this case, researchers reasoned that if both human and primate females are orgasmic, this suggests that the female orgasm has some evolutionary purpose, but as already noted, it's not necessary for a woman to have an orgasm in order to reproduce. What then might the purpose of the female orgasm be? Ask that question in a room full of biologists, and a heated debate is likely to ensue.

On one side of the issue are those who argue that female orgasms might have a variety of evolutionary purposes, such as encouraging women to have sex, playing a role in mate selection, or even sucking sperm into the uterus by means of muscular contractions.

On the other side of the debate are those who maintain that female orgasms actually have no purpose, that they're just a fun and accidental by-product of evolution. The reasoning goes that there's such strong evolutionary pressure for men to have orgasms that women end up getting them too, as a serendipitous bonus, because the necessary nerve cells form before the embryo develops into a male or female. It's analogous to men having nipples, which they have no need for, because of the evolutionary pressure for women to have them. (The analogy to male nipples pops up *a lot* in scientific discussions of female orgasms.) So in this version of events, one could even say that orgasms are the male species' gift to women. The theory may sound a bit chauvinistic, but it has wide support from both male and female biologists.

However, the primatologist Sarah Blaffer Hrdy offers a disturbing possibility that straddles both sides of the debate. She suggests that the female orgasm might once have had a purpose, but it no longer does, because in the modern world a woman's ability to have an orgasm has become entirely detached from her reproductive success, and what evolution doesn't need, it often phases out.

In support of Hrdy's hypothesis, surveys of female sexuality routinely find that a large percentage of women don't have orgasms during sex. So while researchers were arguing about the orgasmic potential of female apes and monkeys, it may have made sense to look a little closer to home. It's possible to imagine a future a million years from now in which all women are non-orgasmic. A rather dystopian scenario, to be sure. Though by that time, if the human race manages to survive, our descendants may have dispensed with their physical bodies entirely and downloaded their brains into vast, computer networks. In which case, orgasms won't be an issue at all.

The Chimp Who Grew Up Human

Norman, Oklahoma – 1972. As he walks up the path towards the ranch-style house, Stephen adjusts his tie and smooths his jacket. He reaches the door, rings the bell confidently, and waits. After a few moments, a bearded man dressed casually in brown corduroys and a short-sleeved shirt opens the door. 'Can I help you?'

'Good afternoon, sir. My name is Stephen. I'm from the American Bible Company. I was wondering if I could take a few minutes of your time to share with you a wonderful opportunity for your family.'

The bearded man looks him up and down critically. 'You're a Bible salesman?'

'Yes, sir. I have some beautiful Bibles I'd like to show you.'

The bearded man thinks for a second, then nods. 'Well, you better come in then.'

He steps aside to let Stephen in the door, leads him through the front hallway into the living room, and gestures towards the sofa. 'Please have a seat. By the way, my name is Maury.' The two men shake hands. Stephen sits down on the sofa, places his black bag in front of him, opens it, and carefully removes a Bible that he sets on the coffee table. The Bible is bound in white leather with gilt edges. Maury sits in a chair opposite him.

The salesman looks directly at Maury and smiles. 'Thank you for inviting me into your home, Maury, and allowing me to share this opportunity with you. I like to think that I have the best job in the world because I'm placing a product in homes that people will never be dissatisfied with.'

As he's speaking, a large hairy figure bounds in from an adjoining room. The salesman glances over, registers the presence of the creature, and recoils with shock. 'Sweet Lord! It's a monkey!'

Maury scowls. 'She's not a monkey. This is my daughter, Lucy. She's a chimpanzee.'

Lucy rushes up to the salesman, puts her furry hands on his knees, sniffs him, and looks directly into his face. She stands about three feet tall, with bent knees. Her fur is deep black interspersed with flecks of chocolatey brown.

Stephen freezes in place. 'Is . . . is she gonna hurt me?'

Maury laughs. 'Don't worry. She's quite friendly. She just wants to say hello.'

Lucy puts her long arms around Stephen's neck, then climbs up onto his lap and kisses his cheek.

Stephen remains rigid. 'She's very large,' he whispers. Lucy starts to run her fingers through his hair and places her mouth over his.

'Lucy,' Maury calls out. 'For heaven's sake, leave the poor man alone! Go get a drink.' Lucy turns around to look at Maury, who makes a gesture with his hands. Lucy signs something back, hops down onto the floor, and walks into the kitchen.

Stephen, looking pale, breathes a sigh of relief. His muscles visibly relax. 'She's quite something. I've never been that close to a chimp before. Kinda took me aback. You say you keep her as a pet?'

'She's my daughter,' Maury repeats.

'Your daughter.' Stephen nods. 'Sure, I understand. I know a guy who calls his cats his children. Does she ever bite?'

'She's entirely harmless, I assure you, unless she were to perceive you as a threat. If that were to happen, she could easily tear you limb from limb. By the way, would you like something to drink? A gin and tonic, perhaps?'

'No, thank you. I don't drink.' At the mention of being ripped limb from limb, Stephen has turned a little paler.

'Well, Lucy is going to have something, but don't feel you need to join her. You see, you arrived during her cocktail hour.'

'Her cocktail hour?'

Stephen turns to watch Lucy in the kitchen. She has a glass in one hand and a bottle of gin in the other. She opens the bottle with her mouth, pours herself a generous shot, then puts the bottle down and reaches into the refrigerator for some tonic water. Her drink prepared, she takes a sip and smacks her lips appreciatively.

'She knows how to make drinks!' Stephen blurts out.

'Oh, she knows a lot more than that!'

'I've been all across the country and seen many things, but I've never seen anything like this!'

Lucy walks back into the living room with her drink, climbs up on the sofa beside Stephen, and sets her glass down on the table.

'Why don't you continue telling me about that Bible,' Maury says.

'Oh, of course.' Stephen looks down at the book in front of him and clears his throat.

'As you may know, the Bible is the number one bestseller in the world.' He stops to watch Lucy again, who's now leaning over the side of the sofa noisily searching through a stack of magazines. She finds what she's looking for and triumphantly holds it up in the air – a copy of Playgirl.

'That's . . . that's . . .'

'Lucy's favourite magazine,' Maury fills in. 'She likes to look at naked men.'

Lucy places the magazine in front of her on the sofa, then leans back and begins flipping through the pages with her feet.

'Just ignore her,' Maury urges. 'Please keep telling me about the Bible. That looks like a beautiful binding.'

'Umm . . .'

Stephen stares transfixed at Lucy, who's making happy grunting sounds, 'Uh, uh, uh, uh,' as she gazes at the naked men. She reaches out and strokes a penis tenderly, then looks coyly up at Stephen. She shifts position on the sofa, moving closer to Stephen, and touches his arm.

'What is she doing?' Stephen asks, a note of alarm in his voice.

'Oh, don't mind her. She's in oestrus right now. You see, she's never had any contact with her own species, so all her sexual interests are directed toward human men.'

As Maury says this, Lucy stands up beside Stephen, puts her arm around his neck, and starts to rhythmically thrust her genitals against his body.

Stephen pushes her away and leaps up. 'GOOD GOD! I've got to be going. Right now!'

He flings the Bible into his bag, tucks the bag beneath his arm, and races towards the door. With the agility of an Olympic athlete, he pulls *it open, and is gone. The screen door swings back and forth from the force of his departure.*

Maury leans back in his chair and watches the salesman disappear down the path. He turns to look at Lucy, who's still standing on the sofa beside where Stephen was sitting. She looks at Maury quizzically, as if to ask, 'Where has he gone?' Maury shrugs his shoulders. 'I don't know, Lucy. But I was just about to buy that Bible!'

At the start of the twentieth century, a bold vision gripped the scientific imagination – to create a humanized ape. Initially, as we've seen, researchers envisioned that such a creature would be ideal for work, being stronger than humans yet capable of performing complex tasks no other animal could master. Techniques for achieving this goal were dreamed up and tested: intensive training, human–ape hybridization, gland therapy . . . But by the 1930s, it had become clear that none of these techniques would work. The apes remained stubbornly ape-like.

So fantasies of creating an ape worker faded away. However, the vision of creating a humanized ape remained as alluring as ever. The motive simply shifted. The goal of creating a worker slave was replaced by the desire to find an intelligent companion.

The new quest to create a humanized ape resembled the search for extraterrestrial intelligence, but turned earthwards and directed into the jungle. Such a creature, if brought into existence, would prove that humans were not the only species capable of rational thought. We would no longer be alone! And it seemed obvious to the promoters of this agenda that a creature of the wild couldn't be moulded into a civilized being by brutal methods of discipline or biological manipulation. Violent tactics would only beget savagery and distrust in kind. Instead, it would require love and nurturing to effect such a transformation. So a new plan formed: to place apes in caring human families and thereby cause them to grow up human.

Civilizing Feral Children and Apes

The idea of using the power of nurturing to transform a wild creature into a human had its roots in the eighteenth century, when 'feral children' found living in the wild attracted the attention of European scholars. Famous cases included Peter of Hanover, a young man captured in a German forest in 1724 who communicated only by gestures and cries, and the Wild Boy of Aveyron, seized near Saint-Sernin-sur-Rance, France, in 1797 as he tried to sneak onto a farm and steal food. These were children who had been abandoned as infants and grew up without significant human contact, but they looked and acted so alien – covered in filth, grunting, scrambling around on all fours – that scholars were at first unsure whether they really were human. In the twelfth edition of his *Systema Naturae*, Carl Linnaeus classified them as a separate species: *Homo ferus*.

Researchers tried to civilize the wild children, but with poor results. Jean Itard, a physician from the Institution Nationale des Sourds-Muets, spent six years struggling to teach the Wild Boy of Aveyron how to interact with other humans before admitting failure. After all that time, the boy still possessed almost no language skills, and his manners left something to be desired. For instance, he often pulled down his pants and defecated in public, or suddenly started masturbating – his favourite activity that engaged him for long hours every day.

Such cases highlighted the importance of a nurturing environment for human development. Humans may possess innate intelligence, but it became clear to psychological researchers that social interaction at a young age was necessary to develop higher critical faculties. If the initial window of nurturing was lost, if an infant had no human contact, then language and social skills never developed and couldn't be taught later. Barring a miracle, the child was doomed to behave, in many ways, more like an animal than a human for the remainder of his or her life.

As twentieth-century researchers studied chimpanzees and

gorillas, it occurred to them that apes might be like feral children. Perhaps apes possessed innate, human-like intelligence, but this intelligence could only develop if it were nurtured in an appropriate environment. Perhaps if apes grew up with humans, they would soon be talking and reasoning like a person. There was only one way to test this theory – place an ape in a human family and see what happened.

The first experimental ape adoptions occurred early in the twentieth century. Nadia Kohts in Russia and Oskar Pfungst in Germany both briefly raised baby apes, although their experiments weren't widely publicized. But it was in America that a mania for home-rearing apes fully took hold. Something about the idea appealed to the American democratic spirit. It played to the conceit that anyone, even a chimp or gorilla, could aspire to become a solid, middle-class citizen.

In 1931, the American researchers Winthrop and Luella Kellogg took into their home a seven-month old female chimp named Gua, on loan from the Yerkes Anthropoid Center, in Orange Park, Florida. The Kelloggs proceeded to raise Gua as if she was their own child, alongside their human son, ten-month-old Donald. Being able to observe a human and ape infant side by side offered a rare chance to learn about the comparative development of the two species, but nine months into the experiment, Gua didn't seem very human. Donald, on the other hand, had started to adopt ape behaviours. When his mother heard him ask for food by imitating his sister's 'food bark', she decided enough was enough and pulled the plug on the trial. The Kelloggs shipped Gua back to the Yerkes Center. She died a year-and-a-half later of fever.

Several other Yerkes researchers later participated in similar home-rearing studies. In 1938, Glen Finch and his family briefly adopted a chimp named Fin, and in 1947, Keith and Cathy Hayes took Viki, a female chimp, into their house. The Hayeses kept Viki for seven years and tried to teach her to speak by moulding her mouth into the shape of words, but they had no more success than

Itard with the Wild Boy of Aveyron. She only learned to voice four words – mama, papa, cup, and up. Other notable non-Yerkes efforts included Maria Hoyt, who in 1932 adopted Toto, a baby gorilla, and Lilo Hess, who in 1950 acquired a young chimpanzee named Christine from a pet shop and raised her on a Pennsylvania farm.

The Norman Experiment

By the early 1960s, there had been a number of home-rearing experiments – enough so that researchers knew that raising an ape among people didn't magically transform the animal into a human. However, this didn't stop scientists from continuing to dream of using nurturing to create a humanized ape. In fact, the most ambitious effort yet was about to occur.

It took place in the sleepy residential town of Norman, Oklahoma. Bill Lemmon, a professor at the University of Oklahoma in Norman, had purchased a farm there in 1957 and was slowly transforming it into a primate research centre. In the early 1960s, he still owned only a few chimps, but his ambition was to turn his farm into a world-class facility where cutting-edge psychological research would be conducted. Ape language experiments and home-rearing studies were central areas of interest for him.

The problem with all previous home-rearing studies, as Lemmon saw it, was a lack of rigour. They had started too late in the ape's life, and had ended too early, after the frustrated experimenters realized that taking care of an ape was more than they had bargained for. But if the experiment was going to be done, Lemmon figured, it should be done right. The chimpanzee should be acquired at birth, separated completely from its own species, and immersed completely in a nurturing human environment. Most important of all was not to give up too soon. It was essential to have patience and to be totally committed to the project. Thus began the story of Lucy, destined to become the most fully humanized chimpanzee ever created.

Lemmon found two colleagues, Maurice and Jane Temerlin,

willing to serve as the host family. Maurice (known to everyone as Maury) was head of the University of Oklahoma's psychology department. Jane was Lemmon's assistant. The Temerlins vowed to commit to the project unreservedly and wholeheartedly. They would treat the chimpanzee, whenever a suitable one was found, as their own child, not as a research subject.

The search then began for a newborn chimp. In 1964, through his contacts, Lemmon heard about a pregnant chimp at a small roadside zoo, Noell's Ark Chimp Farm, in Tarpon Springs, Florida. He hastily made arrangements to acquire the infant, and on the day of the birth Jane Temerlin flew out to Tarpon Springs where she found the mother huddled in a corner of her cage, holding her newborn infant to her breast, suspiciously eyeing anyone who came too close. Jane handed the mother a glass of Coca-Cola spiked with a sedative. The mother took it, sniffed it carefully, and drank it down. After a few minutes, she fell asleep. Jane then climbed into the cage and delicately pried the baby girl from her side. She placed the baby in a bassinet and carried her home to Oklahoma on a commercial airline. The little chimp, soon to be named Lucy, had just taken her first step on the path towards becoming human.

Lucy's Early Life

Maurice Temerlin told the story of Lucy's first ten years in his book *Lucy: Growing Up Human*, published in 1975. The experiment began much like earlier home-rearing studies. As a baby, Lucy charmed everyone. Men and women alike felt an urge to hold, cuddle, and stroke her. The Temerlins stayed true to their word and treated her like their own daughter. They surrounded her with love and never physically punished her. Just like any other child, Lucy ate at the table with the family, had the run of the house, and played with neighbourhood kids. She was even treated by a paediatrician, not a vet.

But, of course, Lucy wasn't a human child, and gradually, as was expected, her chimp nature asserted itself. By a year-and-a-half, she was climbing everywhere, opening cabinets and getting into trouble. The Temerlins put locks on all the cabinets and hid the phones after she placed a call to Tel Aviv.

By the age of three, Lucy could, as Maurice wrote, 'take a normal living room and turn it into pure chaos in less than five minutes'. She unscrewed light bulbs, dragged toilet paper throughout the house, removed books from shelves, and shredded the padding in furniture. It was when Lucy cut herself by punching out a glass window that the Temerlins decided, for her own protection, to modify the experiment. They built an addition onto the house, made out of reinforced concrete with a steel door, where they kept her while they went to work. Nevertheless, the essence of the experiment remained unchanged.

When Lucy was four, a young researcher from Lemmon's lab, Roger Fouts, began making daily visits to teach her American Sign Language. By this time she had begun to display behaviour that was eerily human. Every morning when Fouts arrived, she would hug him and then put a kettle of water on the stove and make him tea.

The inspiration for the sign-language instruction came from another chimp research project, led by Allen and Beatrix Gardner. Observing how unsuccessful the Hayeses had been in their efforts to teach a chimp to vocalize human words, the Gardners reasoned that a chimp's vocal cords simply weren't designed for speech. Chimps, however, are naturally expressive with their hands, and this gave the Gardners the idea of teaching a chimp (Washoe) sign language.

After only a year, the Gardners' results were so promising that Fouts, their student, decided to teach sign language to Lucy also. Soon she had mastered over one hundred words. She even became the first chimp to lie to a human when Fouts discovered one day that she had defecated on the living-room floor. He pointed at the pile of faeces. 'Whose dirty dirty?' he asked. Lucy looked around innocently. 'Sue!' she signed back. Sue was another student whom Lucy knew. Fouts shook his head. 'It not Sue. Whose that?' Lucy looked away. 'Roger!' she signed. Again Fouts shook his head. 'No! Not mine. Whose?' Finally Lucy broke down and admitted the truth. 'Lucy dirty dirty. Sorry Lucy.'

The Experiment Grows Strange

Though the experiment with Lucy began much like earlier home-rearing studies, as it progressed elements of strangeness started to creep in. Truth be told, the strangeness didn't come from Lucy, who simply did what any chimp would do in similar circumstances, but rather from the Temerlins, who, it turned out, were not quite your average middle-class family. They had a passion for flouting convention, which perhaps should have been obvious from their decision to adopt Lucy in the first place.

When Lucy was three, Maurice noticed she was constantly eating rotten apples in the yard. He realized the fermented fruit was making her high, a sensation she seemed to enjoy. So to wean her off the rotten fruit, he gave her an alcoholic drink before dinner. Lucy loved it. Plus, it put her in a good mood. Soon she was having cocktails every day with Maurice and Jane. Her favorite was a gin and tonic in the summer and a whiskey sour in the winter.

Cocktail hour became a carefully observed ritual. If Maurice forgot to prepare her a drink, she initially feigned disinterest, then stole his drink, ran into an adjoining room, and swallowed it down in a gulp. On a typical day, Lucy liked to lie on her back, sipping her drink as she lazily browsed through a magazine. Occasionally she got up and danced around the room. Sometimes she posed in front of a mirror, making silly faces and laughing. Maurice described her as 'an ideal drinking companion' because of her pleasant demeanour while drunk. As he put it, 'She never gets obnoxious, even when smashed to the brink of unconsciousness.' In other words, Lucy had a bit of a drinking problem.

However, it was as Maurice tracked Lucy's sexual development that her life departed fully from mainstream middle-class values. At the start of the experiment, Maurice had decided to keep an eye on what effect Lucy's unique upbringing would have on her sexuality. From a scientific point of view, this curiosity seemed appropriate, and taken in this light, his descriptions of Lucy's first tentative explorations of her body all appeared reasonable: her exploration of her genitals, and even her discovery of the pleasurable sensations imparted by a vacuum cleaner – how she would run the vibrating hose all over her body, 'chuckling with great delight'. Even the five pictures Temerlin provided of Lucy masturbating with a vacuum didn't seem out of place.

But what Temerlin did next proved more controversial and disturbing. He noted that Lucy was totally uninhibited about masturbation. How, he wondered, would she react if he were equally uninhibited? To find out, he pulled down his pants and masturbated in front of her. Lucy was unfazed. Next Maurice arranged for Jane to masturbate in front of Lucy. Again, no response. But when both parents went at it together, Lucy grew highly agitated and grabbed their hands, trying to make them stop. Eventually, Lucy grew so disturbed by the idea of her parents having sex that Maurice had to sleep in a different room. Maurice didn't reveal to readers that there was another reason for the separate sleeping quarters: Jane, by that time, was having an affair with Bill Lemmon. Although Maurice kept this information private, the deteriorating marriage would later have enormous consequences for Lucy.

Life in the Temerlin household grew even weirder as Lucy approached the age of sexual maturity, eight years old for a female chimpanzee. Maurice wondered whether she would seek him out as a sexual partner. After all, she only knew humans, so her sexual desires would be directed towards them. Thankfully, when the time came, she left him alone – though readers were left to wonder how he would have responded if she hadn't. But not only did she not seek him out as a sexual partner, Maurice discovered she displayed a strong aversion to him whenever she went into oestrus. He attributed this to an instinctual incest taboo. 'She will not hug, kiss, caress, or cuddle with me,' he complained. Towards other men, however, she was far more affectionate. 'She made the most blatant and obviously sexual invitations,' Temerlin confided to a *Chicago Tribune* reporter. 'It doesn't matter who – Bible salesmen, Fuller Brush men – she leaps in their arms, covers their mouths with hers, and rhythmically thrusts her pelvis against their bodies. This can be disconcerting.'

To satisfy Lucy's awakening sexual curiosity, Temerlin bought her copies of *Playgirl* magazine, which soon became her favourite reading material. She lay on the floor, stroking the naked men with her finger while making low grunting sounds. She often tried to mock copulate with the centrefold, positioning it below her on the floor while she gyrated up and down on top of it.

But having assured readers that Lucy made no sexual advances towards him, Maurice then shared a disconcerting story that seemed to contradict this statement, dropping it into the text of his book as a casual aside. 'It is curious that I never experienced desire for her,' he wrote, 'even though Lucy would see me nude and attempt to put my penis in her mouth.' The passage continued:

> Although this may appear sexual to the reader, it never felt that way to me. I always felt that her mouthing my penis was exploratory rather than lustful as she never subsequently 'presented' her genitalia and she never once attempted this behavior when she was in estrus. Furthermore, Lucy attempts to mouth my penis whenever she sees it, whether I am urinating, bathing, or have an erection. As a matter of fact, I think it is accurate to say that Lucy is fascinated by the human penis since she attempts to explore it with her mouth whenever she can.

Apparently, Temerlin shared Bill Clinton's definition of what does and does not constitute sexual behaviour. Thanks in large part to self-confessional gems such as these, by 1974 Temerlin found himself without a job at the university. Tenure, he discovered, only protects a professor so far.

What's to Be Done with Lucy?

When Temerlin completed his book, Lucy was ten years old. Most children at that age are looking forward to high school. But Lucy, of course, wasn't like other children, and Temerlin admitted her future was uncertain. Lucy had been home-raised far longer than any other chimp, but Maurice and Jane worried about what would happen if Lucy outlived them. It seemed sensible to arrange a more permanent, stable home for her.

Just as importantly, the two parents were finally wearying of the effort. Maurice had lost his job, and his marriage was on the rocks, two unfortunate developments that life with Lucy had greatly contributed to. Both Maurice and Jane yearned for a 'normal life', but the only way to achieve this was for Lucy to go. The question was, go where?

Every night as they sat at the kitchen table, the couple gloomily surveyed the options. Should they send her to a zoo? Donate her to a research centre? Or find a private ape colony willing to accept her? All of these choices sounded a lot like sending their precious daughter to prison. Even worse, it would be a prison where she wouldn't recognize the other inmates as her own species. Lucy, after all, had still never met another chimp. But then a fourth option swam into view.

In the African nation of Gambia, an experimental chimp rehabilitation centre had recently opened where captive chimps were being taught survival skills and then released back into the wild. It was the first such programme of its kind, and the idea immediately caught the Temerlins' imagination. They envisioned Lucy living free in the jungle, as chimps should live. Impulsively they made the decision: Lucy would go to Africa.

It was true Lucy wouldn't be behind bars, but were they really giving Lucy a chance at freedom, or were they abandoning her in the jungle? Controversy still simmers around that question. Lucy was so fully domesticated that her idea of foraging for food was raiding the refrigerator. She was no more equipped for life in the wild than any pampered, city-living teenage girl would be. Nevertheless, the course was set. Having first been humanized, Lucy was now going to be dehumanized. And so a new chapter began in her life.

In September 1977, the Temerlins and Lucy arrived at Gambia's Abuko Nature Reserve. They sweated in the heat and swatted away the flies. All of them felt uncomfortable. Accompanying them was a young graduate student, Janis Carter, who had previously been Lucy's room cleaner, shovelling up her faeces and scrubbing the floor. The Temerlins hired her to come along and help with the transition. The plan was to spend three weeks with Lucy, settle her into her new home, and then leave.

The Temerlins departed Africa on schedule, but when it came time for Carter to go, she hesitated. She looked at Lucy, who, in that short span of time, had become a pathetic, nervous wreck – scared, confused, and homesick. Lucy wanted her old life back. She had no idea what was happening to her, and Carter realized that if she left, Lucy would, in all probability, soon die. So Carter made a decision that changed her life. She stayed to help Lucy.

Carter moved into a treehouse on the grounds of the nature reserve, where she diligently studied books on primate behaviour – what they ate, where they slept, how they raised their young – in order to teach Lucy how to be a chimp. Lucy, meanwhile, stayed in a protected enclosure, where she pouted, sulked, and did everything in her power to avoid the other chimps, thinking they were strange, unpleasant creatures. She still clung to the hope that she would be taken home soon.

Word got around that Carter was a soft touch who couldn't say no to a sad face, and soon every young orphaned chimp that showed up in nearby towns was being passed along to her. Within a short span of time, she was taking care of nine chimpanzees, including Lucy.

Carter lived in the nature reserve for eighteen months before deciding it was time to leave. She feared that if she stayed any longer she would soon be taking care of every stray chimp in the region. But also, Carter had concluded that Lucy needed a more dramatic break with human civilization, since in the protected environment Lucy was making little progress. It was time for Lucy to move into the jungle, where conditions would force her to behave more like a chimp. So Carter packed all nine of her wards into a Land Rover and together they set off across miles of dusty, dirt roads. The strange troop of chimpanzees, led by a single white woman, eventually ended up at Baboon Island, a fly-infested spit of land in the middle of the Gambia River.

Lucy in the Jungle with Janis

Baboon Island hardly offered luxury accommodations. There was no electricity or running water. Nor were there other people for miles around. The local tribes avoided the place, believing a dragon-like creature haunted it. It was just Carter and her chimps – as well as the indigenous baboons, vervet monkeys, hippos, hyenas, and spitting cobras.

With the help of some British soldiers who were passing through the area, Carter built a wire cage to give herself some protection at night. When the sun set on the first day, she lay down on her cot, exhausted. She was woken by something soft and pungent landing on top of her. Carter realized that the chimps, fearful of the jungle and trying to stay as close to her as possible, had congregated on top of her cage. There they were defecating whenever a stray sound scared them, and the faeces were falling down onto her. Carter immediately resolved that teaching the chimps to sleep in the trees would be a priority.

During the next few weeks, Carter began the process of teaching the chimps to live like chimps. She wandered through the forest with her furry retinue trailing along behind her, attempting to teach them foraging skills. If she spotted a bunch of fresh green leaves, she would imitate a chimpanzee food bark, then greedily devour the leaves. The chimps carefully inspected her mouth to make sure she had really eaten them before sampling themselves. Carter hid her sense of nausea and also ate bugs and biting ants, since these are an important source of protein for wild chimps. The experiment proved to be as much about the chimpification of Janis Carter as it was about the dehumanization of the chimps.

All the chimps made good progress, except Lucy. She had no interest in being a chimp. She wanted to drink water out of a cup like a human, not out of the river. She wanted to sleep in a bed, not in a tree. She didn't want to forage for food. Civilized beings, after all, got their food out of a refrigerator. She would sit on the ground as the other chimps climbed in the trees and wait for stray bits of fruit to drop down. Often she would point at another chimp, or at Carter, and sign, 'MORE FOOD, YOU GO!'

A battle of wills ensued between Lucy and Carter that lasted over a year. Again and again Carter showed Lucy how she needed to behave in order to survive, but Lucy, like a refined lady, stuck her nose up in disgust and turned her back. Eventually Carter grew so frustrated she decided only one option remained to her – tough love. She would withhold all special attention from Lucy until she cooperated.

First Carter withheld food and special attention. When that didn't work, she stopped communicating with Lucy in sign language. In a cruel twist of irony, after all the time people had spent teaching Lucy how to sign, the key to her survival had now become the unlearning of that skill. Faced with the loss of language, Lucy went into a state of shock, as if her world was ending. She sat outside Carter's cage for hours on end, signing: 'Food . . . drink . . . Jan come out . . . Lucy's hurt.' Carter did her best to ignore her.

It got so bad Carter worried Lucy might die. Lucy grew pathetically thin and tore much of her hair out. But finally, as Carter put it, Lucy 'just broke'. After a particularly bad day, the two of them collapsed together on the sand and fell asleep. When they woke, Lucy sat up, picked a leaf and handed it to Carter. Carter chewed it, then handed it back to Lucy, who also ate it. It was the first time Lucy had eaten a leaf. Progress, finally, had been made.

From that day forward, Lucy's behaviour improved. She started foraging and socialized more with the other chimps. She even adopted a baby chimp, Marti. After another year, Carter felt comfortable enough to leave Lucy alone and moved to a camp downstream. At first she returned frequently to check on Lucy's progress, but eventually her visits became less frequent.

In 1986, after being off the island for six months, Carter returned to check up on the chimps. Lucy, hearing the sound of Carter's boat, wandered out of the jungle to greet her. The two hugged and kissed. Carter had brought along some of Lucy's possessions from her life in America, thinking Lucy might enjoy seeing them and that enough time had passed so they wouldn't stir up painful memories. Lucy glanced at them briefly, but then turned away, as if they didn't interest her. Her greeting concluded, she left the objects on the ground and walked back into the jungle. To Carter, this was a sign that Lucy was finally more chimp than human. The transformation back into her own species, though difficult, seemed to have worked.

Unfortunately, Lucy's story didn't have a happy ending. When Carter returned a few months later, in 1987, she couldn't find Lucy. Frantically she organized a search party. Eventually they found the scattered remains of a chimpanzee near Carter's old campsite. The remains had to belong to Lucy, since she was the only chimp missing.

It's not clear how Lucy died. Carter speculated that poachers killed her. Lucy, feeling a bond with humans, might have approached a group of hunters, only to be shot. Or perhaps she was bitten by a snake, or just fell ill. We'll never know for sure. Carter buried the remains beneath Lucy's favourite food tree.

Humanization's End

Lucy's death marked a symbolic end to the century-long effort to humanize apes. It wasn't that her death elicited a great public outcry – the media, for the most part, ignored her passing – but it coincided with a decline in funding for primate behavioural research. In the late 1970s, for instance, Lemmon's facility at the University of

Oklahoma closed. Most of his chimps were sent to a laboratory in New York, where they became subjects in vaccine research.

There were a variety of reasons for the decline. Financial difficulties played a major role. Caring for apes is expensive, and the cost had become prohibitive for many centres. But also, after decades of inconclusive experiments, many people had simply given up hope that primate behavioural and language studies would yield any results. Exacerbating this sense of futility was the dramatic unravelling in 1977 of a high-profile, five-year ape-language study at Cornell University, in which the lead researcher, Herbert Terrace, had come to the conclusion that there was no evidence apes possessed a capacity for linguistic communication similar to a human.

So the twentieth century's quest to humanize our primate cousins ended in failure. We have no ape workers, nor do we have apes living happily in human homes and the one chimp who almost became human lies in Africa in an unmarked grave. May she and the dream of ape humanization both rest in peace.

CHAPTER FIVE

Do-it-Yourselfers

'The experiment will not be easy,' the scientist says to the assembled crowd. 'It will require great sacrifice. You may cry out in pain. You may beg me to stop. But you can rest assured that your suffering will not be in vain. You will be contributing directly to the great cause of the advancement of knowledge. Do I have any volunteers?' The scientist gazes around hopefully at the faces of the men and women. They stare back at him silently. Many of them look down at the floor. There seem to be no takers, but finally a voice cries out, 'I'll do it!' Everyone looks around to identify the brave person. Then they realize it's the scientist himself speaking. The crowd bursts out in applause. What a noble gesture! What a heroic man! The experiment begins, and it's as bad as the scientist warned. He howls in agony. He writhes and contorts. But, oddly, there's a smile on his face. At last it ends. However, the researcher cries out, 'Again! I must have more data.' The entire excruciating ordeal repeats itself. Then he insists on doing it a third and a fourth time. People shift uneasily on their feet. When he cries, 'Again!' a fifth time, they cast worried glances at each other. An uncomfortable suspicion starts to creep into their minds. Is this indefatigable self-experimenter really a hero, or is he simply a madman?

Hard to Swallow

Pavia, Italy – 1777. Warm morning sunbeams shine through the window, illuminating the laboratory of Lazzaro Spallanzani. The forty-eight-year-old professor sits at a wooden table dressed in woollen breeches and a loose white cotton shirt. He's stoutly built, short-necked, high-shouldered, and bald except for a smattering of thinning white hair. Around him the shelves of his laboratory sag with leather-bound volumes of scientific works, specimens of curious creatures in glass bottles, minerals, and taxidermied birds. But Spallanzani focuses his attention solely on a large ceramic bowl that sits on the table in front of him.

He stares pensively at the bowl, as if expecting his breakfast to materialize inside of it. For a few long minutes he doesn't move. Then he sighs deeply, glances upward, utters a short prayer beneath his breath, stands up, and leans over the bowl. Opening his mouth wide, he roughly shoves his finger down his throat.

'Hhuuuuaaarrrrggggg!' He gags violently and his ribs heave. Spasms rack his midsection. Spittle dribbles from his mouth into the bowl.

Spallanzani removes his finger from his mouth, straightens himself back up, and wipes his forehead with his sleeve. His face is red from the strain of retching. 'Mamma mia!' he mutters. Then he leans back over and jams his finger again into his throat.

'Huakkkk huaaakkk uuuuaaaagghhhhh!' His entire body trembles from the force of the gag reflex. With his free hand he grips his straining chest muscles, as if experiencing a heart attack, but he doesn't relent from shoving his finger down his oesophagus. Abruptly a stream of bile gushes up out of his stomach and spills over his hand into the bowl. Still he continues to jam his finger down his throat, until his chest heaves again, and suddenly, choking, he pulls his finger from his mouth and clutches his neck.

Violent, gagging convulsions rack his body. His face, already flushed red, deepens to an angry shade of purple, as if about to burst. Again and again he retches, struggling to free something from his throat. Then, with a final spasm of his chest, an object flies from his mouth and falls into *the bowl with a clatter. Spallanzani gasps for breath and collapses back into his chair.*

After a few moments, he leans forward to inspect the contents of the bowl. A shallow puddle of frothy gastric fluid rests at the bottom of it, and in the fluid floats a small wooden tube, about an inch in length. Spallanzani delicately picks up the tube and holds it up to the light. It glistens with slimy fluid. Inside it he can spy a partially digested morsel of beef, its surface soft and gelatinous.

'Si! Si!' he says. 'Grande!'

He reclines back in his chair, gingerly rubbing his ribcage with his free hand, and with a happy grin on his face examines the meat-filled tube more closely.

Infants possess a powerful urge to explore their surroundings by sticking things in their mouth. Chubby fingers reach out, grasp toys, crayons, or any other small, loose items, and transport them directly into the gaping mouth of the child. There the tongue carefully tastes and feels them, and then, much to the concern of anxious parents, often swallows them. The mouth is so full of nerve endings that it makes sense for infants to acquire information about the world in this way, but by the age of three, oral exploration typically gives way to safer visual and tactile scrutiny. Children lift up objects, squint at them, turn them around in their hands, and then fling them across the room, laughing. By the time we're adults, most of us have become extremely conservative about what we're willing to put in our mouths. We might taste a new wine, or sample some foreign food, but that's usually the limit of our courage.

But what's true for the average person isn't necessarily true for scientists. Many researchers over the years have rediscovered the wisdom of infants. They've realized that their mouths (and stomachs) can be extremely useful tools for discovery. As a result, there's a robust tradition among self-experimenters of swallowing stuff for the sake of science. All manner of objects – edible, nonedible, poisonous, pathogenic – have disappeared down the throats of researchers.

Swallowing Tubes

The study of digestion provided the earliest motive for scientists to swallow foreign objects. In the late eighteenth century, doctors had little understanding of how the body digested food. The leading theories were that food either fermented, rotted, or was crushed to little pieces within the stomach. To shed some light on this mysterious subject, the Italian scientist Lazzaro Spallanzani came up with the idea of placing small pieces of food inside tiny linen bags, sewing them shut, swallowing them, and inspecting what emerged at the other end. This method allowed him to trace individual food items as they journeyed through his gastrointestinal tract.

Spallanzani started his digestion experiments in 1776, while teaching at the University of Pavia. The first food he tested was a chunk of bread that he briefly chewed before spitting it into a bag. The thought of the bag getting stuck in his gut, with painful, possibly fatal consequences, caused him some apprehension, but then he remembered that people swallow indigestible items such as cherry stones and plum pits every day with no ill effect, and this reassured him. So he placed the bag in his mouth and washed it down with a large glass of water. Then he nervously waited to see what would happen. The next morning he had a bowel movement, and there in his chamber pot, much to his pleasure, was the linen bag. He hadn't experienced any discomfort at all. He fished out the bag, cleaned it off, and opened it up. The bread was gone! The speed of the process suggested that the bread had neither fermented nor rotted, but instead had been dissolved through the cloth by gastric juices.

More tests followed. Spallanzani enclosed bread in double and triple layers of cloth bags. Protected by three layers, a small morsel of bread survived a trip through his stomach. Next he replaced the bread with meats – pigeon, veal, beef, and capon. If he pre-chewed the meats, his body digested them. Otherwise they came out partially digested. Curious to see how long it would take to finish digesting unchewed meat, he swallowed a semi-digested piece a second time. Part of it still remained after re-emerging, but a third pass through his system did the trick. He didn't specify whether he placed the meat in a clean bag each time.

To test whether the squeezing of stomach muscles aided digestion, Spallanzani screwed up his courage, abandoned the linen bag, and stuffed some chewed veal inside a small hollowed-out, perforated wooden tube, which he then swallowed. Again it passed through him without a problem, dropping into his chamber pot empty of meat. This revealed that the stomach doesn't digest food by grinding or crushing it.

Before concluding his investigations, Spallanzani wanted to examine gastric fluid more closely. He also hoped to prove that digestion occurs primarily in the stomach, not the intestines. He addressed both issues simultaneously by swallowing a tube containing some beef. Then he waited three hours, stuck his finger down his throat, and vomited the tube back up, along with a large quantity of gastric fluid.

Some people, such as supermodels, can make themselves vomit quite easily, but Spallanzani apparently didn't possess this skill. He strained and suffered before the tube finally re-emerged from his stomach. But once he had the tube back in his hand, he noted that the beef had already turned soft and gelatinous, which proved that digestion had been occurring in the stomach. Next he immersed a fresh piece of beef in some of the gastric juice and sealed it inside a glass bottle. Three days later the meat had dissolved into an oozy slime, whereas a piece of beef kept for a similar length of time in water had begun to rot, but was still whole. This conclusively demonstrated that digestion was different than decomposition, and took place far more quickly.

Spallanzani would have liked to conduct more experiments with gastric fluid, but he couldn't handle the idea of further vomiting. He explained:

> The disagreeable feelings occasioned by the act of vomiting, the convulsions of my whole frame, and more especially of my

stomach, that continued for several hours after it, left upon my mind such a repugnance for the operation, that I was absolutely incapable of repeating it, notwithstanding my earnest desire of procuring more gastric liquor.

Thanks to Spallanzani's efforts, scientists learned that digestion is a chemical process. Coincidentally, he wasn't the only researcher studying digestion at the time. Almost simultaneously, a young medical student in Edinburgh, Edward Stevens, conducted a similar set of experiments, unaware of Spallanzani's research. However, Stevens didn't use himself as a subject. Instead he convinced a mentally handicapped Hungarian street performer to serve as his guinea pig. Stevens described him as a man 'of weak understanding who gained a miserable livelihood by swallowing stones for the amusement of the common people at the imminent hazard of his life'. A highlight of the man's act was that the stones could be heard clunking against each other as they navigated the circuitous route of his intestine.

Stevens persuaded this man to put his talents to better use by repeatedly swallowing a perforated silver ball that Stevens loaded with various test materials – raw beef, fish, pork, cheese, apples, turnips, wheat, barley, rye, and mutton bone. The results he obtained were similar to those of Spallanzani, but Stevens concluded his investigation by asking a curious question. He wanted to find out if a living creature could survive a journey through a man's gut. So he placed a live leech into the silver globe, reasoning that leeches are exceptionally tough parasites. It emerged from the Hungarian's posterior transformed into a 'black viscid miasma', which answered his question in the negative. An earthworm that Stevens made the Hungarian swallow suffered a similar fate. The experiments ended when the performer abruptly left Edinburgh, presumably to find a new audience for his act elsewhere.

The Man Who Ate Glass

In 1916, 140 years after Stevens' Edinburgh experiments, a skinny, nervous man walked into the University of Chicago laboratory of digestion specialist Anton Julius Carlson and introduced himself as Frederick Hoelzel. Like Stevens' Hungarian performer, Hoelzel possessed the peculiar ability to eat just about anything. Reporters eventually nicknamed him the 'Human Billy Goat'. But, unlike the Hungarian, Hoelzel was eager to use his talent for the sake of science, and he eventually became a researcher in his own right.

Hoelzel's interest in digestion began in 1907, at the age of eighteen, when he suffered a series of severe stomach aches. He felt that he never regained the health he had enjoyed before this illness, and he grew convinced that food itself was the source of his problems. He was sure that if only he could starve himself, he would feel well again. So he embarked on a series of amateur self-experiments to discover a non-caloric substance that would curb his hunger and allow him to reduce his food intake. However, he had difficulty finding anything that met his needs. Charcoal, the first thing he tried, wasn't filling enough. Next he ate sand seasoned with salt. It reduced his hunger, but irritated his intestines. It was like defecating sandpaper. Rounded glass beads seemed to roll out of his stomach as fast as he could swallow them, leaving him feeling hungrier than before.

Hoelzel continued to sample a huge variety of materials – loam, corn cobs, sawdust, nut shells, cork, feathers, hair, wool, sponge, straw, rubber, asbestos, chalk, silk, flax, rayon, and banana stems – but for various reasons he crossed each one off his list. Finally he tried surgical cotton cut up into small pieces. It wasn't perfect, having the tendency to cause anal leakage and rectal itching, but he considered it the best of all the substances he had tested, and it became a staple part of his diet. He initially accustomed himself to it by drizzling maple syrup on it, but eventually he taught himself to eat it plain. As a treat he liked to dip it in orange juice.

Hoelzel eagerly wrote to researchers, peppering them with questions and describing his theories about digestion and nutrition, but most doctors rebuffed him. Hoelzel later frankly admitted, 'My sanity was questioned in some medical quarters.' He might have continued conducting his amateur self-experiments for the rest of his life, ignored by the scientific community, but Carlson saw some potential in him. He took Hoelzel under his wing, encouraged him to continue his education, and steered him towards more rigorously conducted scientific research. The two men often collaborated. Hoelzel never earned the title of doctor, but he became a frequent contributor to journals such as *Science* and the *American Journal of Physiology*.

The first experiment Hoelzel undertook with Carlson, in 1917, was a fifteen-day fast. Carlson wanted to find out whether the sensation of hunger would disappear after a few days without food, as had been anecdotally reported. A photograph of Hoelzel in his underwear, taken immediately before the fast began, shows him looking very thin and somewhat downcast. In a second photograph, taken fifteen days later, he's even skinnier – skeletally thin; his underwear hangs baggily off his emaciated frame – but now a slight, satisfied smile curls the edge of his lips. Hoelzel could have probably continued the fast longer, but after fifteen days it had become apparent that his hunger hadn't disappeared, so Carlson's question had been answered. When he resumed eating, Hoelzel wrote glumly in his diary, 'Food does not taste as good as I anticipated.' In fact, the need to eat never seemed to afford him any pleasure. He was happiest when starving himself. Later, Hoelzel endured fasts as long as forty-two days.

In 1924, Hoelzel undertook a more unusual series of investigations. During five weeks of daily tests, he ate a variety of inert materials – rock, metal, glass – and measured their rate of passage through his gastrointestinal tract. Glass beads proved to be the speediest of all the substances, travelling through his alimentary canal in a mere forty hours, on average. Next came gravel, scooped up from a walkway outside the lab, which rattled out into the toilet fifty-two hours after he ate it. Steel ball bearings and bent pieces of silver wire each took approximately eighty hours to pass through him. Gold pellets were the slowest, moving at a leisurely pace through his intestines, and only emerging after twenty-two days.

Eating inert materials subsequently became part of Hoelzel's daily routine. Between 1925 and 1928 he ate up to one hundred pieces of knotted twine every day, keeping an obsessive record of his bowel movements in order to record the twine's rate of passage. His intestinal speed record was achieved by one knot that zipped through his system in a mere one-and-a-half hours, aided along by a violent bout of diarrhoea. Other regular meals included rubber tubing flavoured with spaghetti sauce and steel bolts and nuts. One time an assistant in Carlson's lab served him a small monkey wrench fried in batter as a joke. It wasn't recorded whether Hoelzel ate it. In 1929, Hoelzel switched to consuming five grams of metal pellets daily. Christmas was the only day of the year he took a break from this grim fare, allowing himself a small, but plain meal of digestible food.

By 1930, Hoelzel had become a full-time guinea pig, living a monk-like existence in a small room attached to Carlson's lab. He received no wages, but was given his room and board (such as it was) in return for his services. He doesn't seem to have had much of a life outside of the lab. For recreation he took short walks in the nearby park. Occasionally he attended motion pictures alone in the evenings. In scientific publications he described himself as an 'Assistant in Physiology' at the University of Chicago.

The extreme diet and lifestyle took its toll on Hoelzel. An unnamed reporter who visited Carlson's lab in 1933 was shocked by Hoelzel's appearance, writing in an article distributed by *American Weekly, Inc.*: 'His hands are like those of an invalid, white, blue-linen and bony, his Adam's apple stands out from a scrawny neck, and his skin is colourless except for a network of fine blue lines, especially under his eyes.'

Nevertheless, Hoelzel remained convinced of the dangers posed by food, and of the benefits of not eating. In 1947, he co-authored a paper with Carlson demonstrating that periods of intermittent fasting appeared to prolong the life span of rats – the link between caloric restriction and longevity remains a subject of intense scientific scrutiny to this day. And as late as 1954, Hoelzel wrote that he regarded overeating to be 'the number one problem of civilization'. However, Hoelzel's minimal diet didn't noticeably prolong his own life. He died in 1963 at the age of seventy-three.

A Diet of Worms

Hoelzel's diet of cotton, metal, gravel, and glass is hardly likely to attract many converts, but his choice of edibles may seem downright appetizing when compared to what researchers in some other fields have chosen to ingest. Helminthologists – that is, researchers who study parasitic worms – stand out as particularly heroic diners.

Parasitic worms were one of the first disease agents identified by medical practitioners. References to them can be found in ancient Egyptian papyri, and scholars believe they're also described in the Bible – some historians argue that the 'fiery serpents' that afflicted the Israelites after their exodus from Egypt were actually guinea worms that grow subcutaneously and emerge through the skin of their host. However, it wasn't until the nineteenth century that researchers began to fully unravel the life cycle of parasitic worms and made the first attempts to prove that the small white larvae found in many raw meats could, if eaten, grow into worms inside the intestines.

One of the first attempts to experimentally infect a human host with worms occurred in November 1859, when the German scientist Gottlob Friedrich Heinrich Küchenmeister fed a condemned prisoner raw pork riddled with the larvae of *Taenia solium*, a common tapeworm. An executioner decapitated the prisoner on 31 March 1860, after which Küchenmeister eagerly cut into the man's body to uncover what might have grown inside of him. The physician found almost twenty worms wriggling inside the intestines. Küchenmeister considered the worms to be comparatively small, although the largest one measured five feet. The tenacity with which the worms clung to the lining of the dead man's gut impressed him. When he tried to remove some of them, they forcefully attempted to reattach themselves.

Küchenmeister was widely criticized for experimenting on a prisoner, but he defended himself by noting that pomegranate extract was known to be an effective treatment for tapeworms. Therefore, in the unlikely event that the man had been pardoned, his intestinal infestation wouldn't have doomed him to a lifetime of suffering. Even so, the prisoner apparently wasn't given a choice about becoming a worm incubator.

The distinction of being the first researcher to experimentally infect himself with worm larvae went to the Sicilian professor Giovanni Battista Grassi. On 10 October 1878, while conducting an autopsy, Grassi discovered the large intestine of the corpse to be full of the larvae of *Ascaris lumbricoides*, a giant intestinal round-worm. It occurred to him that this find, foul as it might have been, represented a novel opportunity for an experiment. He could eat the larvae to prove that ingestion was the source of infection!

Grassi wanted to make sure he conducted the test properly. This meant that first he had to prove he wasn't already infected. So he fished the larvae out of the dead man's intestines and placed them in a solution of moist excrement, where he could keep them alive indefinitely. Then he microscopically examined his own faeces every day for almost a year to confirm his lack of infection. Finally, on 20 July 1879, he felt confident he was free of worms, so he spooned some of the larvae out of their faecal home and ate his unpalatable meal. A month later, to his pleasure, he experienced intestinal discomfort, and then found *Ascaris* eggs in his stool. Having confirmed his infestation, he treated himself with male fern, a herbal anti-worm medicine, and flushed the immature parasites out of his intestines.

After Grassi, it became increasingly popular for helmintholo-gists to eat worm larvae and grow parasites inside their intestines. In fact, it almost came to be seen as a gruesome rite of passage within the profession. From Japan, in 1886, came the report of Isao Ijima, who proudly reported that he grew a 10-foot *Bothriocephalus latus* worm in his gut. The next year, Friedrich Zschokke and ten of his students at the University of Basel in Switzerland downed tapeworm eggs found in local fish. A few weeks later, after taking anti-worm medication, they removed worms up to six feet in length from their intestines. Their actions received widespread coverage in the media, inspiring people throughout Europe to send them letters of praise and encouragement. Only one correspondent was critical – a man from the local chapter of the Society for the Prevention of Cruelty to Animals who criticized them for conducting such a dangerous experiment.

By the twentieth century, it had become fairly common for researchers to eat worm larvae. Nevertheless, the actions of the Japanese paediatrician Shimesu Koino still impressed the scientific community and continue to be spoken of reverentially by helminthologists to this day. In 1922, Koino ingested a record-breaking 2,000 mature *Ascaris lumbricoides* eggs. A month later, after developing severe flu-like symptoms, Koino found *Ascaris* larvae in his sputum. In this way, he deduced the curiously roundabout life cycle of the creature in humans – that after being ingested, the larval worms migrate from the intestine into the blood and thereby end up in the lungs. From there, they're coughed up, reswallowed, and return to the small intestine, where they finally grow into full reproductive adults.

American researchers also partook of parasitic delicacies, proving that worm larvae were a truly international cuisine. In 1928, Emmett Price of the US Bureau of Animal Husbandry swallowed the larvae of an unknown parasite he found in the liver of a dead giraffe at the National Zoo in Washington, DC. Newspapers gleefully reported Price's unappetizing self-experiment, but the media seemed even more amazed when Price's boss casually explained that he not only approved of what Price had done, but had expected him to do it, since it was a tradition within the parasitology section of the bureau for researchers to use their own bodies as worm-growing labs. One reporter exclaimed:

He had to do it because it is the 'code of the bureau'. A harsh code that requires a man to make a test tube out of his own stomach and a laboratory of his own body. There are no exact conclusions that may be drawn from the heroic action of this scientist, except that it shows how very far removed from the average man are the glories of the scientific man. . . . Most people will draw the line at becoming private feeding grounds for the parasites of giraffe livers.

This peculiar habit of helminthologists has endured right up to recent times. In 1984, Soviet researcher V. S. Kirichek sampled tapeworm eggs he found in the brains of reindeer living in the far north of Russia, and in Taiwan, in 1988, researchers at the Department of Parasitology in Taipei's National Yang-Ming Medical College reported consuming tapeworm larvae found in the liver of a local Holstein calf.

The Vomit Drinkers

There are worse things than worm larvae to consume – far worse things. Eating vomit, for instance, might provoke greater revulsion. Eating the black, blood-streaked vomit spewed up by victims of yellow fever would presumably inspire even more intense repugnance. And yet there are researchers who have dived willingly into these depths of disgust. During the first half of the nineteenth century, it actually became something of a fad for doctors to drink fresh, steaming black vomit.

The Philadelphia doctor Isaac Cathrall pioneered the stomach-turning practice of vomit drinking. After attending medical college in Europe, Cathrall returned in 1794 to Philadelphia, where he found a yellow-fever epidemic ravaging the city. At the time, yellow fever was one of the most feared diseases in the world since it was a swift, brutal, seemingly unstoppable killer.

Yellow fever victims first develop a sudden and severe headache, followed by fever, acute pain in their muscles and joints, weakness, and inflammation of the eyes and tongue. As the liver grows diseased, the skin acquires a greenish-yellow tinge – thus the name of the disease. Finally, blood often starts leaking into the gastrointestinal tract, where it coagulates, causing the victims to produce copious amounts of a black vomit that resembles slime-soaked coffee grounds. Death can follow a mere three days after the first display of symptoms.

To most laymen at the end of the eighteenth century, it seemed obvious that yellow fever was contagious since after one case showed up in a region the disease usually spread rapidly throughout the population. Therefore, people shunned the sick. However, many doctors had begun to suspect that yellow fever wasn't contagious, as direct contact with the victims – even exposure to the stinking discharge of their bodies – didn't seem to result in illness. The germ theory of disease hadn't yet been developed. Instead, doctors theorized that foul or polluted air, exacerbated by sweltering summer heat, caused the outbreaks.

As Cathrall tended to the sick in Philadelphia, he became convinced that yellow fever wasn't contagious. He grew so certain of this belief that he decided to put it to dramatic test by designing a series of experiments to demonstrate the benign effect of black vomit on a healthy subject: himself. He began in October 1794 by removing a quantity of black vomit from the stomach of a dead man. He then liberally applied this to his lips and licked it with his tongue. He wrote, 'It gave, a short time after application, the sensation of a fluid perceptibly acrid.' Decades later, another Philadelphia doctor, René La Roche, concurred with this description of the taste of black vomit, characterizing it as bitter, acidic, and 'more or less insipid'.

Next Cathrall smeared black vomit all over his skin and immersed his hand in a bucketful of vomit freshly ejected by a patient. Again, he felt nothing unusual. He fed the vomit to cats, dogs, and fowl. None of them showed signs of ill health. Finally he sat for an hour in a room suffused by the noxious vapours of heated black vomit. He wrote:

> The fluid was evaporated until the atmosphere was so impregnated with the effluvia of the vomit, as to render the apartment

extremely unpleasant, not only from the odour of the vomit, but the warmth of the room. In this atmosphere, I remained one hour; during which, I had a constant propensity to cough, and had, at times, nausea and inclination to vomit; but, after walking out in the air, these effects gradually subsided.

Despite all these efforts, Cathrall didn't grow ill. Several years later, in 1802, a young Philadelphia medical student named Stubbins Ffirth continued Cathrall's investigations. Like Cathrall, he smeared black vomit on himself, but then he went further. He placed black vomit in his eyes and swallowed an undiluted cup of it. He too suffered no illness, which seemed to prove the non-contagion theory rather conclusively.

Soon the drinking of black vomit spread beyond Philadelphia. In 1816, the French doctor Nicholas Chervin imbibed large amounts of the substance while studying the disease in the West Indies. But it was another French doctor, Jean-Louis-Geneviève Guyon, who went furthest of all, and arguably conducted the most nauseating self-experiment ever in the history of science, in his passion to prove, beyond a shadow of a doubt, the non-contagious nature of yellow fever.

While at Fort Royal, Martinique, in 1822, Guyon first took the sweat-drenched shirt off a yellow-fever victim and wore it for twenty-four hours. Simultaneously he removed yellow discharge from a festering blister on the patient and rubbed it into cuts on his own arm. After waiting a few days, Guyon announced he felt no fever. But this test merely served as a preliminary run for what followed.

Guyon next dressed himself in the blood-and-sweat-soaked shirt of a man who had just died of yellow fever. Because Guyon was present at the death, he was able to remove the shirt immediately after the man took his last breath, so that it still retained the heat of the fever-racked body. Then Guyon lay down in the excrement-soiled bed. He rolled around in the filth, smearing the diarrhoeal faeces over his skin to ensure maximum exposure and remained there for six hours. He rubbed 'black sanguineous matter' extracted from the stomach of the patient onto his arm. Finally, to top it all off, he drank a cup of the man's black vomit. Then he waited to get sick. Miraculously, he remained healthy.

Guyon may have gone to the most extreme lengths to prove the non-contagious nature of yellow fever, but the man who drank vomit with the most style was the British naval surgeon Robert McKinnel. While McKinnel was aboard the HMS *Sybille* in 1830, yellow fever broke out on the ship. Being quite certain the disease was caused, as he wrote, by 'noxious emanations from the interior of the ship', McKinnel calmed the fears of the crew by filling a wine glass with the black vomit of a sufferer. He raised the glass in the air and offered a toast to a fellow officer: 'Here's to your health, Green.' Then he drank the glass down. He reported that it didn't even impair his appetite for dinner.

All these doctors who suppressed their own sense of nausea and determinedly drank down cups of black vomit have been, for the most part, forgotten by medical historians. This is because, despite their bravery, the vomit drinkers were wrong. Yellow fever definitely is contagious, although they were partially correct in that it's very difficult to catch the disease by contact with a victim. Instead, it's transmitted almost exclusively by the bite of mosquitoes. The Cuban physician Carlos Finlay first suggested the role played by mosquitoes in 1881. A team of researchers led by Walter Reed confirmed Finlay's theory in 1900. One of Reed's researchers, Jesse Lazear, sacrificed his life to prove the hypothesis by purposefully exposing himself to the bite of a mosquito that he knew carried infected blood. It was an action as bold and dramatic as anything Cathrall, Ffirth, or Guyon did, but not nearly as revolting.

Filth Parties

Drinking various substances to prove their pathogenic, or non-pathogenic, nature became a fairly common form of self-experimentation after the development of the germ theory of disease in the latter half of the nineteenth century. For instance, in

October 1892, Max von Pettenkofer drank bouillon laced with *Vibrio cholerae* to prove his theory that the bacterium was not the primary cause of cholera. Pettenkofer was wrong, but luckily he suffered only a mild case of diarrhoea as a result of his error. More recently, in 1984, the Australian researcher Barry Marshall ingested water infected with *Helicobacter pylori* to prove that this bacterium was the cause of gastritis and stomach ulcers. Two weeks later he developed stomach discomfort, nausea, and bad breath, indicating his theory was correct. Marshall later won a Nobel Prize for his discovery.

However, only one series of self-experiments during the last century-and-a-half has come close to matching the unsavoury extremes of the vomit drinkers. These were the so-called 'filth parties' of Joseph Goldberger.

In 1914, the US Public Health Service assigned Goldberger to lead a commission to study pellagra, a potentially fatal disease that was reaching epidemic proportions. The disorder is character-ized by irritated skin, diarrhoea, and dementia. Goldberger initially thought, as did virtually the entire medical community, that pellagra was an infectious disease. But the more he studied it, the more convinced he became that the condition was actually caused by a dietary deficiency. To prove his theory, Goldberger took a cue from the earlier yellow-fever researchers and designed a series of experiments in which he, and a team of sixteen other volunteers, attempted to infect themselves with pellagra in order to demon-strate that it was impossible to do so.

The self-experimenters collected skin extracts, nasal drippings, urine, and faeces from pellagra victims. They mixed these speci-mens with wheat flour and cracker crumbs, rolled them into a pill shape, and ate them. As the experimenters sat around sharing these excretory snacks, they jokingly referred to their gatherings as 'filth parties'. Some of the volunteers experienced nausea and diarrhoea, but none of them came down with pellagra.

Other researchers subsequently confirmed Goldberger's suspi-cion, determining that pellagra was caused by a chronic dietary deficiency of niacin (also known as vitamin B3). So as a result of

Goldberger's investigations – and his willingness to eat urine and faeces – a disease that once afflicted millions of people has now almost entirely disappeared in the developed world.

A disturbing thought to end on is that the ingestion of scatological specimens is unfortunately not confined to the realm of self-experimentation. All of us may be unwitting participants in filth parties whenever we choose to dine out. Study after study has shown restaurants to be awash in germs. A recent investigation, published in 2010 in the *International Journal of Food Microbiology*, fingered the self-serve soda dispensers found at many fast-food restaurants as particularly egregious founts of foulness. People handle the dispensers after going to the bathroom, and the dispensers then pump the faecal contamination and bacteria directly into the carbonated drinks. The levels of contamination are low enough to not pose a serious health risk, but the thought alone of what you're consuming might make you sick!

This Will Be Extremely Painful

Fayetteville, Arkansas – 10 July 1922. 'She's an excellent specimen,' *William Baerg says, peering into the glass jar at the black widow spider, his eyes gleaming with excitement. Baerg is in his late thirties with dark hair and an athletic build. Beside him, a young college student named Garlington nervously eyes the spider. They're seated at a bench in a laboratory filled with jars containing exotic-looking insects of various kinds including scorpions, centipedes, and cockroaches. An aquarium in the corner houses a large brown tarantula.*

'I last fed her over forty-eight hours ago,' Baerg continues. 'So she should be ready to bite.' He picks up the jar and unscrews the lid.

'Professor, are you sure you want to do this?' Garlington asks. 'Couldn't we test it on another rat, or a cat?'

'That wouldn't tell us the effects on man. Anyway, I'm sure it won't

be that bad. Just make sure you take good notes. I don't want to have to do it again!' Baerg laughs, and Garlington smiles awkwardly.

Using a pair of forceps, Baerg reaches into the jar and gently grabs the spider. He lifts it out and holds it up for view. It's about an inch in diameter, shiny black, with a large globular abdomen and a dark-red hourglass marking on its underside. 'Beautiful, isn't she!' he exclaims. Garlington nods but flinches back.

'So let's proceed. Are you ready?'

Garlington opens the journal on the bench in front of him and picks up a pen. 'I'm ready.'

'OK. Let's see.' Baerg glances at the clock on the wall. 'It's 8.25 a.m. I am now going to attempt to make the spider bite me.' Garlington obediently transcribes the information into the notebook.

Baerg delicately removes the spider from the forceps, gripping it by its abdomen between the thumb and first two fingers of his right hand. 'It's actually not that easy to get these little fellows to bite. There's a trick to it.' Garlington keeps writing as Baerg extends his left hand forward and places the spider's head against his index finger. 'You want to place the fangs against the selected spot and then gently move the spider from side to side just so.' His actions match his words. 'Hopefully one of the fangs will catch in the skin and this will induce the spider to implant both of them as deeply as possible.' As if on cue, the spider sinks its fangs into his skin.

'There we go. Perfect!' Baerg smiles broadly. 'The sensation is rather faint, like the prick of a sharp needle. Ah, now the pain is increasing in intensity.'

A brief look of horror flickers across Garlington's face as he stares at the spider hanging onto the professor's finger. Then he directs his attention back to his journal.

'Yes, it's quite painful now. Sharp and piercing. OK, it's been five seconds. I think that's enough.' He pulls the spider away from his finger, quickly drops it back in the jar and seals the lid.

Garlington visibly relaxes as soon as the spider is again contained behind glass. Baerg stares at his finger. 'Where her fangs penetrated, the flesh looks slightly white. There's a small drop of clear fluid, but otherwise

I can't see any puncture mark. However, the pain hasn't lessened at all. In fact, it seems to be growing in strength and spreading throughout my finger.'

Garlington stops writing and looks at the professor.

'Well, this should be an interesting experience,' Baerg continues, shaking his finger to lessen the pain. 'I hope I get a full reaction!' He grins happily. Garlington purses his lips with concern, hesitantly nods his head as if not quite sure he agrees, and then, once again, picks up his pen and starts to write.

Physical hardship is sometimes an unavoidable part of scientific research. Scientists may have to collect data in remote, inhospitable settings, such as Antarctica or the top of a volcano. They may have to handle caustic materials or risk infection. Researchers usually take these difficulties in their stride. They accept discomfort as an occasional but necessary part of their job.

However, there are cases in which scientists don't simply accept discomfort, but actively seek it out. Through self-experimentation, some researchers repeatedly inflict pain on themselves, as if purposefully testing the limits of their endurance. Like the religious ascetics of old who practiced mortification of the flesh – whipping themselves with thorny branches or praying for hours while kneeling on freezing-cold flagstones – they punish their bodies, refusing to be deterred by the screaming agony of their nerve endings. These are the self-experiments that blur the line between scientific curiosity and masochistic desire.

A Protopathic Sensibility

The physiological study of pain offers ample opportunities for any scientist with a penchant for self-punishment. And in fact, there's a tradition within the field that those who research pain should have first-hand knowledge of its effects. The doctor responsible for establishing this tradition was the pioneering neurologist Sir Henry Head.

While working at London Hospital in the early twentieth century, Head began a study of patients who had suffered nerve injuries, hoping that such injuries would provide clues about the manner in which the nervous system functioned. However, he soon grew frustrated. The patients, lacking his training, couldn't provide the kind of detailed responses he wanted. They were only reliable, he commented, for the 'simplest introspection' – that is, yes and no answers to his queries. He concluded that if he wanted better information he was going to have to rely on himself. So he decided to transform his body into a living laboratory of pain by giving himself a nerve injury.

Head persuaded a surgeon at the hospital to perform the necessary operation. On 25 April 1903, Dr Henry Dean made a six-and-a-half-inch incision in Head's left forearm, folded back a flap of the skin, severed the radial and external cutaneous nerves, rejoined the nerves with silk sutures, and then sewed Head back up. A large region on the back of Head's hand promptly went numb.

Injuring himself was the easy part. The more difficult process was tracking and analysing his recovery. To help with this, Head enlisted the aid of the Cambridge neurologist William Rivers. Every weekend for the next four years, Head travelled from London up to Cambridge, where Rivers subjected his arm to a battery of tests. Rivers systematically poked pins into Head's arm, touched glass tubes full of scalding hot water to it, pulled the hairs, rubbed it vigorously with cotton wool, and even froze areas of the flesh with ethyl chloride. As this was going on, Head sat with his eyes closed, his chin resting on his right hand, and told Rivers everything he felt, or didn't feel.

From a scientific perspective, Head's sacrifice was worth it, because the two researchers made a major discovery. After seven weeks, a rough form of sensation returned to Head's hand, allowing him to sense temperature and pain, but it was almost two years before he regained the ability to feel more fine-tuned sensations, such as light touches. Based on this, the researchers concluded, rightly, that the body has two distinct sets of nerve pathways, which they called the protopathic and epicritic. The protopathic is the body's system of first response. It alerts the body to pain and temperature, but in a vague, diffuse way. The epicritic system gathers more precise, detailed tactile information.

Curious about whether any part of the normal skin exhibited purely protopathic characteristics, Rivers carefully examined Head's entire body. Sure enough, he found what appeared to be such a region – the tip of the penis.

To explore this discovery further, the two men decided additional experiments were in order. Rivers took his colleague's penis in his hand and poked it roughly with stiff fibres, producing pain that was 'so excessively unpleasant that H. cried out and started away'. Rivers then dipped the penis into increasingly hot glasses of water as Head stood with his eyes closed. This part of the experiment turned out to be not entirely disagreeable. Head reported that when the water temperature reached 45° Celsius the feeling was initially painful, but soon yielded to an 'exquisitely pleasant sensation'. Finally, Rivers pressed rods of varying diameters against the penis tip and asked Head to tell him the size of the rods. Head could feel the pressure but had no idea about the relative size of the object poking him. Based on these observations, the researchers concluded that the tip of the penis was indeed 'an organ endowed with protopathic and deep sensibility only'. This discovery explains why men can use their fingers, but not their penis, to read Braille.

A Map of Pain

Severing a nerve in one's arm is an extreme act – more than most people would do for their job – but Head didn't appear to have suffered excessively. He never writhed on the floor in agony. So on a scale of pain his effort would merit only two stars out of four. During the 1930s, two other British physicians, Sir Thomas Lewis and his student Jonas 'Yonky' Kellgren, took the agony levels up a notch in a series of self-experiments that explored the phenomenon of referred pain.

Medical researchers first described referred pain in the 1880s, when they noticed that injury to internal organs often causes pain elsewhere in the body. For instance, a disturbance of the stomach might produce tenderness along the ribs. A well-known example of the phenomenon is the pain in the arm that often accompanies a heart attack. Doctors are still not entirely sure what causes the referral of pain.

Lewis and Kellgren decided to produce a comprehensive map of referred pain by systematically injecting saline solution into their muscles, tendons, and joints. The saline was harmless, but it produced intense bursts of pain lasting for up to five minutes before subsiding. They carried on these experiments for years, day after day pushing needles deep into their flesh and recording where on their body this caused pain and tenderness to appear.

For instance, they injected solution into all the ligaments of their spinal column, from the cervical region (neck) down to the lumbar (lower back). One of them would slide the needle into the other's back until the point bumped up against the tough interspinous ligament. Then he would wiggle the point slightly from side to side before injecting the solution. This typically caused an aching hurt to radiate throughout the limbs and chest. They noted happily that 'repeated injections of the same ligament gave remarkably constant results'. Similarly, they plunged needles into each other's buttocks:

> Moving the needle point about vigorously gave rise to a very slight diffuse pain felt in most of the buttock. 0.2 c.c. of 6% saline was then injected into the muscle. This gave a diffuse pain of greater severity felt clearly in the lower part of the buttock and the back of the thigh, and occasionally as far distant as the knee.

Testing bones was more of a challenge. First they tried sticking a needle firmly into bone, but this produced only an unpleasant sensation of pressure. So Kellgren volunteered to have a sharpened, stainless-steel wire driven through his shinbone. He offered the following account of the experience, in which by his matter-of-fact tone he managed to make it sound like having a wire hammered through his leg really wasn't a big deal:

> While the wire was passing through the compact bone I experienced a sensation of pressure and vibration but no pain, but when the wire entered the soft cancellous bone diffuse pain was added to the sensation of vibration. The wire was then replaced by a hypodermic needle and 0.1 c.c. of 6% saline was injected into the cancellous bone. This also gave rise to slight diffuse pain felt widely in the outer side of the leg.

Lewis and Kellgren's self-experimentation, like that of Head before them, yielded practical results. Doctors are now aware, for instance, that if a patient comes in complaining of an aching hip, the knee might be the true source of the pain, not the hip.

The Sensitive Testis

Lewis and Kellgren's experiments required enormous stamina and courage. However, there were even more disturbing zones of discomfort for pain researchers to investigate. It was again a pair of British researchers, Herbert Woollard and Edward Carmichael, who took it upon themselves to explore these extremes. (The British, for some reason, seem to have been particularly enamoured of this kind of research.)

In an article published in 1933 in the medical journal *Brain*, Woollard and Carmichael described the concept of their experiment: 'It had occurred to us that the testis might be regarded as a proper viscus and one suitable upon which observations in regard to referred pain might be made.' Translated into plainer language, this meant they planned to find out what it would feel like to have their testicles crushed.

The two researchers didn't reveal which one of them served as the subject (that is, who had his testicles compressed) and which was the observer, but they did offer specific details about the experimental method:

> The testis was drawn forwards in the scrotal sac and supported by fingers placed below it. A scale pan was rested on the testis and weights placed in this compressed the testis and epididymis between the supporting fingers and the scale pan. Known weights were placed in the pan and left there till the subject described what sensations he experienced and where he felt them.

The researchers conducted five variations of the experiment, using local anaesthetic to numb various nerves leading to the testes in order to examine the role of each nerve in transmitting testicular pain.

Woollard and Carmichael described their results in an extremely dry, detached style, as if purposefully trying to avoid sensationalizing the experiment. One imagines the subject lying spreadeagled on a table, gritting his teeth in pain as his colleague stooped over his groin to stack the pile of weights higher, but no such details can be found in their article. Instead, they recorded only rather colourless clinical observations. For instance, they provided the following results for experiment #1, in which they compressed the subject's right testicle after numbing the posterior scrotal nerve:

300 grm:	Slight discomfort in the right groin.
350 grm:	More discomfort in the right groin . . .
550 grm:	Definite testicular pain followed by a dull ache in the right lumbar region dorsally.
600 grm:	Severe pain on the inner side of the right thigh with indefinite testicular sensation.
650 grm:	Severe testicular pain on the right side.

Readers hear the direct voice of the suffering subject only a single time in the article. During experiment #4, while having 825 grams of pressure applied to his right testicle, the authors record that the subject suddenly exclaimed, 'That is quite different from the left side.' Then the man falls silent again.

Thanks to Woollard and Carmichael's effort, science now knows that approximately one pound of pressure on a testicle will send pain throbbing up to the lower back. Two pounds is enough to make a man hurt all the way up to his upper back. But the most interesting fact they discovered was that even when they used local anaesthetic to block all the nerves leading to the testicles, they couldn't entirely abolish the pain of compression. The testes are extremely sensitive organs!

No other accounts of self-experimentation with testicular pain appear in the medical literature. The only other experiment that even comes close to being similar took place almost forty years later, when two researchers at the University of Texas Medical School, D. F. Peterson and A. M. Brown, decided that the pain caused by testicular pressure was poorly understood and deserved more attention. However, Peterson and Brown didn't use themselves as subjects. Instead they used six male (lightly anaesthetized) cats. They created what they described as a 'homemade supporting device' that cradled a cat's testicle in a cup-shaped depression, and then they pressed down on the tender organ with a metal rod. They learned, to no one's surprise, that cats, like humans, really don't enjoy having their testicles compressed.

Rapid Human Deceleration

Pain researchers have no monopoly on masochistic self-experimentation. Other areas of science certainly offer opportunities for self-punishment. In fact, what perhaps the most celebrated series of physically brutal self-experiments occurred in the field of Air Force medicine, when an American flight surgeon named Dr John Paul Stapp took it upon himself to discover exactly how much abuse a pilot's body could withstand.

Stapp got his first taste of suffering for science in 1946, at the age of thirty-six, when he volunteered for a high-altitude survival experiment. This involved shivering unprotected in a plane at an altitude of 46,000 feet to find out whether deadly gas bubbles would form in his blood. He survived and, in 1951, participated in wind-blast experiments that had him riding in a jet fighter going 570 miles per hour, with the canopy removed. The goal was to determine the maximum wind blast a pilot could endure before being pinned helplessly against the jet. Again he suffered no serious injuries, but the pounding his body endured paled in comparison to the most notorious of his self-experiments – the human deceleration project.

After World War II, the United States Air Force needed to know if pilots could eject from supersonic jets without facing certain death because of the shock of rapidly decelerating from the speed of sound to a near standstill. The transition exposed pilots to forces of over 40 or 50 Gs. (One G equals the force of gravity at the surface of the earth; 40 Gs is like a 7,000-pound elephant falling on top of you.) Many doctors believed that 18 Gs was the most a human body could endure, but no one knew for sure. Stapp volunteered to find out.

At Holloman Air Force Base in New Mexico, Stapp designed a rocket-powered sled that blasted down a 3,500-foot track at speeds up to 750 mph. At the end of the track scoops dug into a pool of water, jerking the sled to an abrupt halt. It went from 750 mph to zero in one second. Early unmanned tests weren't encouraging. Initially the sled skidded off the tracks. Later, the force of the deceleration caused a dummy to break free of its harness and catapult 700 feet through the air. This was followed by a gruesome mishap in which the researchers accidentally brought a chimpanzee to a 270 G stop, instantly transforming it into the equivalent of meat jelly splattered across the desert. But once Stapp felt such kinks had been ironed out, he decided it was his turn to go.

For his inaugural rocket sled ride, in 1947, Stapp went at a gentle 90 mph. The next day he advanced to 200 mph. True to form, he kept signing up for more rides, upping his speed, probing the limits of human endurance. Over a period of seven years he rode the sled twenty-nine times. He took his final ride on 10 December 1954, on which day nine rockets propelled him to 632 mph, faster than a .45 calibre bullet. He outran a jet flying overhead. When the sled hit the water Stapp experienced a record-breaking 46.2 Gs of force, the equivalent of almost four tons of weight slamming down on him.

Each time Stapp rode the sled, the force of the deceleration hammered his body. Stapp repeatedly endured blackouts, concussions, splitting headaches, cracked ribs, dislocated shoulders, and broken bones, but he kept volunteering for more. One time, in a show of bravado, he set a broken wrist himself as he waited for medics to arrive.

The greatest danger was to his eyes. Rapid deceleration causes the blood to pool with great force in the eyes, bursting capillaries and potentially tearing retinas. Even more disturbingly, when a human body comes to a stop that abruptly, there's a real possibility the eyeballs will simply keep going – popping out of the skull and flying onwards. During Stapp's final rocket sled ride, this almost happened. He wrote, 'It felt as though my eyes were being pulled out of my head. . . . I lifted my eyelids with my fingers, but I couldn't see a thing.' He feared he'd permanently lost his vision, but luckily his eyesight gradually returned over the next few days. However, he suffered vision problems for the rest of his life.

Though Stapp suffered greatly, he proved that the human body can endure G forces far higher than previously believed. He suspected that the upper limit lay far beyond the 46.2 Gs he had experienced. This information revealed that crash fatalities were usually caused by inadequate safety equipment, not by the impact of the crash. As a result, the Air Force made numerous improvements to the design of fighter jets, including stronger safety harnesses and reinforced cockpits. Later in life, Stapp zealously promoted automobile safety, and his advocacy contributed to laws making seat belts mandatory in cars. Despite the abuse he put his body through, Stapp lived to the ripe old age of 89.

Neither pain researchers nor Air Force flight surgeons can be accused of shying away from physical trauma. However, the award for the most masochistic self-experiments ever has to go to another set of scientists, the toxinologists. Researchers in this relatively obscure field study the toxins produced by living organisms such as spiders, snakes, bees, ants, and jellyfish. Evolutionary forces have crafted these toxins over millions of years to produce maximum levels of pain – horrific, nightmarish anguish, like having your entire body forcibly turned inside out and dipped in acid. To quite a few toxinologists the existence of such agony-inducing poisons seems to pose a personal challenge. What would it feel like, they wonder, to experience that kind of suffering? In seeking to answer this question, these researchers truly earn the distinction of being the Masters of Pain.

University of Arkansas professor William J. Baerg led the way in this torturous pursuit of knowledge. (Oddly, the 'J' in his name didn't stand for anything. He just thought it looked impressive.) He began his self-experimentation in August 1921, when investigating the effects of the bite of the tarantula, *Eurypelma steindachneri* Ausserer.

Tarantulas, being large, hairy, and scary-looking, are widely feared. Many people assured Baerg their bite was deadly, but he had his doubts. Tests on a guinea pig and rat confirmed his suspicions. Their bite seemed to cause the animals only mild, temporary discomfort, scarcely worse than a bee sting. To be certain, he induced a large female tarantula to sink its black, chitinous fangs into his finger. He described the feeling as being like the stab of a pin, but within two hours the pain had subsided. He concluded that most tarantulas pose little danger to humans.

Emboldened by this experience, Baerg moved on to a spider with an even more fearsome reputation – the black widow, *Latrodectus mactans*. Again, Baerg acknowledged reports of the terrifying effects of its bite. Black widows like to lurk in dark places, such as outhouses. There was one disturbing report, from 1915, of a man bitten on the genitals while sitting on an outdoor toilet. He managed to stagger a mile to the nearest doctor, by which time his penis had swollen to three inches in diameter. However, Baerg noted that such reports were entirely circumstantial. The victims, in their agony, hadn't thought to bring the spider with them to the hospital. So it was impossible to say with certainty that the black widow was the cause of their suffering.

Baerg induced a black widow to bite a rat and found that the effect was, as he put it, 'relatively insignificant'. The rat curled up in its cage, evidently in misery. Occasionally it jerked forwards, as if experiencing convulsions. But within ten hours it had fully recovered. So once again, Baerg decided to test the bite on himself. Unlike the tarantula, however, the black widow packed a punch that fully lived up to its reputation.

On 10 July 1922, at 8.25 a.m., Baerg placed a mature, female black widow on the index finger of his left hand. It quickly sank its fangs into him. He allowed it to bite him for five seconds before pulling it away. Observing all this, and taking notes, was a young student from the university, A. R. Garlington. A sharp, unrelenting pain immediately spread through Baerg's finger. Within fifteen minutes, the ache moved up to his shoulder. Within two hours, it had spread through his arms and down to his hips. Four hours later, he was experiencing painful cramps throughout his entire body. It became difficult for him to speak or breathe.

The rats had recovered from the effects of the bite within ten hours, but nine hours into his ordeal, Baerg's agony was steadily increasing. In addition to the violent cramps and the struggle to breathe, he had developed uncontrollable tremors. Sweat poured out of him. His face contorted in a mask of pain. 'Get me to the hospital!' he gasped.

The attending physician at the hospital, Dr E. F. Ellis, rushed Baerg into a hot bath, which temporarily helped, but soon the symptoms returned full force. As his pain grew in intensity during the night, Ellis made a desperate attempt to draw the poison from Baerg's hand, which didn't work. Ellis also insisted Baerg keep his hand beneath an electric oven, as hot as he could bear, hoping this would ease the pain, but it only made it worse. Eventually Baerg rebelled and refused to keep his hand in the oven.

Baerg didn't sleep during the night. He tossed and turned, feverish, writhing in torment. The next day he felt a bit better, but the respite didn't last because on the second night the suffering returned, with the added bonus of hallucinations. 'Though I slept for short periods,' he wrote, 'I was so delirious that as soon as I would doze I would be frantically and in an utterly aimless fashion working with – spiders.'

On the third day, though he still felt wretched, he could tell the worst had passed. He spent the day in bed, reading the novel *Cytherea* by Joseph Hergesheimer, which he pronounced to be 'nearly as unpleasant as the effects of the spider poison'. That night he was finally able to get some sleep. By the fourth day he was almost back to normal, and Dr Ellis sent him home. A persistent itch bothered him for a few more days, but otherwise he suffered no lasting ill effects.

The severity of the reaction might have persuaded a less intrepid researcher that it had been a mistake to conduct the experiment, but Baerg felt no such regrets. On the contrary, he insisted, 'The unpleasant features were many times compensated for by the fact that I had satisfied my curiosity.' Within a few months, he was back to self-experimentation, determined to find out if the bite of the male black widow was as bad as that of the female. However, the male, being much smaller than the female, couldn't manage to puncture his skin. 'All the response that he made was an indifferent nibbling,' Baerg reported.

In the following years, Baerg continued to match himself against stinging insects, including centipedes and scorpions. Centipedes, he found, 'have a somewhat painful bite, and hold on like grim death', but their poison was extremely mild. Similarly, the effect of a scorpion bite was 'mostly local and similar to that of a wasp or hornet'. Nothing matched the power of the black widow. However, Baerg noted that his observations only pertained to North American insects. Central America hosted varieties of arthropods so deadly that even he wasn't brave enough to test their sting.

Other Stingers

Baerg was merely the first in a series of sting-seeking toxinologists. Other standouts in this pursuit of pain have included University of Alabama professor Allan Walker Blair, who, in November 1933, repeated Baerg's black widow experiment – upping the ante by doubling the dose of venom. Whereas Baerg only allowed the spider to bite him for five seconds, Blair kept its fangs in his finger for a full ten seconds.

Blair had to be rushed to the local hospital two hours later. The attending physician, Dr J. M. Forney, was stunned by his condition. Forney later wrote, 'I found him in excruciating pain, gasping for breath. . . . I do not recall having seen more abject pain manifested in any other medical or surgical condition. All the evidences of profound medical shock were present.'

Despite the agony he was in, Blair insisted the hospital take electrocardiograms to test the effect of the venom on his heart. Blair told the hospital staff it felt like torture to lie still as they made the electrocardiogram, but he forced himself to suffer through it, and the EKG proved to be normal. He remained in the hospital for a week, at one point growing so delirious that he feared he was losing his mind. Like Baerg, after recovering enough to return home, he itched all over for several more weeks.

A curious toxinology self-experiment took place in Australia during the early 1960s after Queensland residents began coming down with a mysterious ailment characterized by painful abdominal cramping, backache, vomiting, and a feeling of sluggishness. The physician Jack Barnes investigated and determined the illnesses were probably caused by a jellyfish sting, though what jellyfish it might be was unknown. After much searching, he finally located a likely culprit, a tiny carybdeid medusa lurking in shallow water near the shore. To make sure the creature really was to blame, he promptly applied its stingers not only to himself, but also to his nine-year-old son. Sometimes children have to pay a heavy price for being the offspring of a mad scientist!

Within minutes, both of them were doubled over in pain. Their cramping muscles caused them to adopt a stance that Barnes colourfully described as being like 'that of an infant with a full nappy'. When they arrived at the hospital half an hour later, they were shivering violently, coughing, retching, and struggling to breathe. Thankfully an injection of pethidine hydrochloride brought almost immediate relief.

The current reigning 'King of Stings' (as dubbed by *Outside* magazine) is Justin Schmidt, research director of the Southwest Biological Institute in Arizona. He boasts that he's been stung by 150 different insect species worldwide, though his speciality is Hymenoptera, the insect order that includes bees, wasps, and ants.

In 1984, Schmidt created the Schmidt Sting Pain Index, which was an effort to classify the painfulness of Hymenopteran stings on a four-point scale. The scale is as follows (the descriptive phrases are Schmidt's own):

0: Unable to penetrate the human skin
1: A tiny spark
2: Like a match head that flips off and burns on your skin
3: Like ripping muscles and tendons
4: You might as well just lie down and scream

Relying on personal experience to collect most of the data for his index, Schmidt has ranked the stings of seventy-eight species of bees, wasps, and ants. The common honeybee comes in at a two. Several species of wasp rank a four, but topping the list, with a four plus, is an ant species, *Paraponera clavata*, also known as the bullet ant because being bitten by one feels like being struck by a bullet. Schmidt had the misfortune to be bitten by three of them in Brazil. He immediately suffered peristaltic waves of pure pain that left him, as he put it, 'quivering and still screaming' hours later.

Schmidt designed his pain index to rank Hymenopteran stings, but there's no reason it couldn't be applied to other forms of pain – spider bites, rapid deceleration, saline injections, testicular compression. For most normal people, such an index would provide a guide to experiences they should avoid. For all the aspiring masochistic scientists out there, however, it might just be a source of inspiration.

Adventures in Self-Surgery

Kane Summit Hospital, Kane, Pennsylvania – 15 February 1921.
Evan O'Neill Kane lies on the table in the operating room. He's a trim sixty years old, bald-headed, wears round wire-framed glasses, and sports a white goatee. His abdomen is bared in preparation for the removal of his appendix. Doctors and nurses dressed in masks, caps, and surgical gowns hover around, ready to begin the operation. They're waiting for the anaesthetist, Theresa McGregor, to finish preparing her equipment.

Kane props himself up on his elbows. 'Wait a second, Miss McGregor,' he says. 'Change of plans. I'm going to perform the operation myself.'

The anaesthetist looks up. 'Sir?'

'I'm doing it myself,' Kane repeats. 'Nurse, put some pillows under my shoulders to prop me up, then hand me a scalpel.'

Kane's brother Tom, the surgeon who was about to perform the operation, steps forward. 'Evan, what are you talking about? Lie down and behave.'

'Sorry, Tom. I changed my mind. Nurse, continue as you were instructed.'

The nurse looks anxiously at the two brothers, uncertain what to do.

Tom shakes his head. 'Evan, don't be crazy. You're not well. You can't remove your own appendix. Lie back and let me do it.'

'I'm not debating this, Tom. I'm the chief surgeon at this hospital, so I outrank you. Nurse, if you want to keep your job, get some pillows and prop me up.'

The nurse obediently nods her head and hurries off to find pillows. Tom stares at his brother, his eyes showing his confusion. 'Evan, of all the boneheaded things you've ever done, this tops them all. I'm not going to stand here and watch as you slice yourself open.'

'Then don't watch,' Evan replies. 'I don't need your help. I can do it on my own.'

'At least you should have warned us. We're not prepared. What if something goes wrong?'

Evan ignores his brother. 'Miss McGregor, please prepare local anaesthetic.'

The nurse returns with pillows that she places beneath his shoulders, propping him up to allow him a clear view of his lower stomach. The anaesthetist hands him a syringe, and without pause he injects himself in the abdomen – several shallow shots and three deeper ones into the abdominal wall. Tom storms off to the far side of the room where the other doctors stand in a group. They look at him sympathetically and shrug their shoulders as if to say, 'We know it's crazy, but what can we do?'

After four minutes, Evan takes a scalpel from the nurse and makes a finger-length incision in his abdomen.

'I can't believe this is happening!' Tom yells from across the room. Unfazed, Evan deftly continues his work. With a few swift strokes he slices through the superficial tissue until he reaches the peritoneum, the membrane lining the abdominal cavity. With scissors and a surgical knife he carefully cuts through it.

'Nurse, my nose itches!' he announces. The nurse reaches over and scratches the tip of his nose.

Evan reaches into his abdomen, lifts up the caecal pouch attached to the intestines, and reveals the worm-like appendix. It's inflamed and enlarged. 'My, my. What do we have here? This looks about ready to burst. You should see the concretions in the distal third!' He looks across the room. 'Dr Vogan, if you would. Could you hold the caecum for me?' With a few steps Dr Vogan is at his side and delicately grasps the organ, allowing Evan to focus on the appendix. Evan shifts slightly, leaning further forward so he can peer deeper into his abdomen. Abruptly he stops moving. 'Oops.'

'What?' his brother shouts.

In a calm voice, ignoring his brother, Evan looks up. 'Nurse, if I could have your assistance, please. My guts appear to be spilling out.'

Medical journals are full of disturbing cases of self-surgery. A quick search turns up a list that reads like something out of a horror novel – self-performed caesarian sections, gouged-out eyes, amputated limbs, castration, trepanation of the skull. And that just scratches the surface. These are the acts of psychotics, religious fanatics, malingerers, the suicidal, sexual fetishists, the feeble-minded, sufferers of excruciating pain, and people in life-and-death situations faced with no other choice. In other words, they're acts of the desperate and insane. Against this tableau of human misery, it's hard to imagine that self-surgery could also be an act undertaken by a qualified surgeon motivated by scientific curiosity, and yet there is a little-known tradition of auto-surgery within the surgical profession, though admittedly it includes only a handful of practitioners.

The Misery of Bladder Stones

The tradition of professional self-surgery belongs almost entirely to the post-anaesthesia period of medicine. During the long, painful centuries before anaesthesia – the majority of human history – self-surgery was the exclusive domain of the desperate and insane. Only when a surgeon had joined their ranks would he dare turn the knife upon himself. In fact, there's only one verifiable case of a doctor operating on himself before the advent of anaesthesia. It was a cringe-inducing attempt to self-remove a bladder stone.

Bladder stones are crystalline masses that form in the urinary tract where they can cause excruciating pain. During the last hundred years they've become a relatively rare condition, but their burning misery used to be well known, probably because poor diet and dehydration from diarrhoeal diseases encouraged their growth.

The only treatment for the condition was an operation known as 'cutting for the stone'. It was the gruesome stuff of nightmares.

A patient lay on his back with his knees raised towards his head. Three or four strong men held him in this position. A surgeon inserted his fingers into the patient's rectum to feel for the stone and hold it in place through the surrounding tissue. Then he made an incision near the anus – without anaesthesia, remember! – and inserted forceps up into the body, pierced the bladder, and removed the stone. Those who didn't die from infection were usually left incontinent and impotent, but at least the painful stone was gone.

Given that this was the only treatment available, some sufferers chose to attempt the operation themselves, in the belief, perhaps, that they had gentler hands. There was a celebrated case of an Amsterdam blacksmith, Jan de Doot, who in 1651 used a kitchen knife to carve a stone the size of a hen's egg out of his bladder. The urologist Leonard Murphy has questioned the details of the story, noting that such a feat seems 'beyond the bounds of possibility and of human endurance'. He argues that the stone probably had migrated out of de Doot's bladder and into his superficial tissues, making it easier to remove. Jan had undergone stone removal (not self-performed) on two prior occasions, and the old wounds could have provided a route for the stone to travel along. Nevertheless, the self-operation still would have required a remarkable effort of will to endure the pain.

The one example from this period of self-surgery by a medical man is a case similar to de Doot's. The optimistically named Clever de Maldigny was an assistant surgeon in the French Royal Guard. In 1824, he found himself afflicted with a stone and decided to do a de Doot. He placed himself before a mirror, firmly grasped a surgical knife, and began to carve away. Eventually he removed a stone that he claimed was 'the size of a large walnut'. As with de Doot, this wasn't his first stone. He had undergone the procedure a mind-numbing five previous times, leaving a well-travelled route for his own surgical intervention. In fact, he discovered that his new stone had formed around a piece of surgical sponge left behind during a previous operation.

There's a rumour that the seventeenth-century physician

William Harvey, who first described the circulatory system, also cut into himself to remove a bladder stone. However, his biographers dismiss the tale, arguing it's more likely he simply passed a large stone in his urine. This leaves Clever as the sole self-cutting, pre-anaesthesia doctor.

The Discovery of Anaesthesia

Anaesthesia finally appeared on the medical scene in the 1840s. Although the authorities now tell us that recreational drug use can only lead to dire consequences, the history of the discovery of anaesthesia offers a case in which the opposite was clearly true, since it was thanks to the widespread use of ether and nitrous oxide as a means to get high that several American doctors independently realized the drugs might also be used to block pain during surgery. Exactly which doctor deserved credit for the discovery became a source of bitter contention.

Crawford Long was a leading contender for the title. He was a rakish young Georgia doctor who liked inhaling ether at parties, finding it a great icebreaker, especially with southern belles. During such a party, he noticed that people high on ether often fell down and received nasty bruises, but they felt nothing until the drug wore off. This gave him the idea of using the drug in his medical practice, which he did for the first time on 30 March 1842, becoming the first doctor ever to anaesthetize a patient. However, he didn't publish his finding – apparently he didn't think it was a very big deal at the time – so he lost the chance to secure credit as the discoverer.

Two years later, a New England dentist, Horace Wells, had the same realization after seeing a demonstration of the effects of nitrous oxide (better known as laughing gas). As an experiment, Wells arranged for a colleague to knock him out with nitrous oxide and then remove one of his teeth. When he woke later, Wells enthusiastically reported he 'didn't feel as much as the prick of a pin'. But again, Wells didn't manage to spread the word of his discovery. He did attempt to stage a public demonstration in 1845, but he didn't give the patient enough gas and the man cried out in pain as Wells removed his tooth. Wells left the surgical theatre in disgrace as the crowd jeered 'Humbug!' after him.

It fell to William Morton, a former colleague of Wells, to convincingly alert the medical community. Morton gave a successful demonstration of the use of anaesthesia during surgery at Massachusetts General Hospital on 16 October 1846. Morton knocked out a patient with ether, and Dr John Warren then removed a tumor from the man's jaw. After the surgery, the patient said he hadn't felt a thing. Warren turned to address the audience: 'Gentlemen, this is no humbug!' The date went down in history as the day surgery changed forever.

Once Long, Wells, and Morton realized the importance of the discovery of anaesthesia, they all scrambled to take credit for it. The result was a noisy ruckus of claims and counterclaims, backstabbing, and argument. Petitioners urged the US Congress to weigh in on the matter and settle it, but it declined to do so.

Wells grew depressed by the vicious politicking. He gave up dentistry and became a chloroform salesman – and addict. One night in January 1848, while living in New York, he inhaled a particularly large dose of the drug and grew delirious. In his madness, he grabbed a vial of sulphuric acid that happened to be sitting on his mantel, staggered out into the streets, and started throwing the acid at prostitutes. A police officer hauled him off to prison. When Wells came to his senses and realized what he'd done, he despaired. He doped himself up with more chloroform – for some reason, the arresting officer had allowed him to bring his medical supplies with him into the cell – and then cut deeply into his leg, slicing through his femoral artery. He bled to death.

Wells was trying to commit suicide, not conduct a scientific experiment. Nevertheless, his actions made him the first doctor to ever perform self-surgery with the aid of anaesthesia, although this wasn't a distinction he'd been hoping for. If he'd lived a few days longer, he would have received the news that the Paris Medical

Society had voted to recognize him as the official discoverer of anaesthesia, which might have lifted his spirits.

The Wonder Drug Cocaine

For fairly obvious reasons, general anaesthesia doesn't lend itself to self-surgery. Wells is the only example of a medical man operating on himself under its influence. It took the discovery of local anaesthesia, almost forty years later, before doctors felt confident enough to cut themselves open for the sake of science. The story of this breakthrough began in 1883, when a young Sigmund Freud decided to self-experiment with cocaine.

At the age of twenty-seven, Freud was a physician at the Vienna General Hospital. His fame as a psychoanalyst lay years in the future. In fact, he was filled with doubt about the direction of his career, and in his uncertainty he cast about for something that might make his name. The answer to his problem, he decided, was cocaine.

People in South America had known for centuries that coca leaves, when chewed, had a stimulating effect. However, scientists only succeeded in isolating cocaine, the active ingredient in the leaves, in 1855. By 1883 the drug was still little known and little understood, but intriguing reports about it were beginning to appear in medical journals. An Italian physician wrote that it helped his digestion. A Bavarian doctor described how, when he gave the drug to soldiers on manoeuvres, it gave them increased energy and endurance.

When Freud read these reports, it occurred to him there might be a medical use for cocaine. If he could just find out what that use was, it would give his career the boost it needed. So he obtained a supply of the drug from his local pharmacy and set out to determine, through self-experimentation, exactly what its medical benefits might be.

Freud mixed up a liquid solution of the drug, and then he carefully examined it. It had, he noted, a 'rather viscous, somewhat opalescent' colour. The smell was aromatic and fragrant. Finally, he lifted the solution to his lips and drank it down. The taste was bitter, but all at once he felt a surge of exhilaration. This was followed by a feeling of lightness and a furriness on his lips and tongue.

Freud loved cocaine and began taking it regularly. He became a cocaine evangelist, excitedly talking up its virtues to anyone who would listen – friends, family members, colleagues. He wrote an article, 'Über Coca', suggesting it could treat ailments as diverse as stomach ache, heart disease and nervous exhaustion. In June 1884, on the eve of this article's publication in the *Archiv Für Der Gesämmte Therapie*, Freud departed for a three-month stay with his fiancée who lived outside of Hamburg, a day's travel from Vienna by train. He wrote to her before he left:

Woe to you, my princess, when I come. I will kiss you quite red and feed you till you are plump. And if you are forward you shall see who is the stronger, a little girl who doesn't eat enough or a big strong man with cocaine in his body.

Freud anticipated that when he returned to Vienna he might be famous, hailed as the champion of cocaine. When he came home in September, cocaine had, in fact, become the talk of the medical community, but much to Freud's frustration, it wasn't because of his article. During his absence, one of his colleagues at the hospital, the ophthalmologist Karl Koller, had also started experimenting with cocaine, having been turned on to it by Freud. Koller quickly zeroed in on the drug's true medical potential. He noted that when people drank it their lips and tongue became numb. So perhaps, he reasoned, it could be used as a local anaesthetic. He tested this by putting drops of it in his eye and lightly poking his cornea with a needle. He felt nothing.

Koller wrote a short paper detailing his discovery, which was read at the 15 September meeting of the Heidelberg Ophthalmological Society. The audience immediately recognized the enormous medical significance of the finding. General anaesthetics were a great benefit to surgery, but they were also dangerous and unpredictable. Too much of them caused serious complications, such as organ damage or even death. There were many times when it would be preferable to avoid those risks and numb only a specific part of the body. Koller's discovery finally gave them that option.

Freud seethed with jealousy as physicians around the world hailed Koller as a hero. He was gentlemanly enough to acknowledge that he had failed to see cocaine's true potential, but to the end of his life he blamed his fiancée (later his wife) for having distracted him by causing him to go on vacation at such a pivotal moment.

Cocaine lent itself to self-experimentation, and following Koller's discovery other researchers enthusiastically used their own bodies to explore the drug's full anaesthetic potential. Fifty-two-year-old New York ophthalmologist Hermann Knapp carefully applied it to various parts of his body to test its numbing effects. He sprayed it on his ears, tongue, up his nose, and down his throat. He squirted it into his urethra using a Eustachian catheter and a balloon. He injected it into his penis with a syringe. Finally, as he reported in an article in the *Archives of Ophthalmology*, 'For the sake of completeness I injected cocaine also into the rectum.' He was pleased to find that he experienced a loss of sensation everywhere he applied the drug.

In Germany, in 1886, Dr August Bier discovered it was possible to inject cocaine directly into the spinal column, thereby blocking sensation in the lower half of the body. He enlisted the aid of his assistant, August Hildebrandt, to explore the full effects of this procedure. The two men stripped naked and performed the spinal injection on each other. As their lower halves grew numb, they squeezed and pinched each other's legs. There was no sensation. Bier stuck a needle in Hildebrandt's thigh, and then poked it deeper until it touched the bone. Hildebrandt shrugged casually. Bier burnt his colleague's leg with a cigar and then struck his shins roughly with a hammer. Still nothing. Finally Bier plucked hairs from Hildebrandt's pubic area and roughly yanked on the other man's testicles. Hildebrandt felt only a slight pressure. After the drug wore off, the two men shared a meal and smoked cigars, entirely satisfied with the results of the experiment.

The Advent of Auto-Surgery

Soon after the discovery of the anaesthetizing properties of cocaine, medical men first used the drug to perform minor operations on themselves. For instance, in 1890, Paul Reclus, a surgeon at the Hôpital Broussais in Paris, developed a tumor on the index finger of his right hand. He consulted with a colleague, who declared that the entire digit needed to be amputated. Reclus, fearing this would end his career as a surgeon, opted for another course of action. He injected cocaine around the tumor and, holding the surgical knife in his left hand, dug out the growth. He reported feeling only a slight queasiness in his stomach as he did so. This was caused not by pain but by the sound of the knife scraping against the tissue around the bone. A few years later, an unnamed sanatorium doctor used cocaine to cut out his ingrown toenails. A Turkish medical student at Paris's Val-de-Grâce removed a mass of varicose veins from his scrotum by similar means.

In 1909, experimental self-surgery finally arrived. Beginning in this year, and for the next four decades, a handful of surgeons performed major operations on themselves – either appendectomies or herniotomies. These weren't acts of desperation. In every case there were other surgeons available and willing to do the job. In fact, the colleagues of the intrepid self-experimenters often begged them not to do it, but the do-it-yourselfers felt the risk would be worth it for the sake of knowledge.

Two questions motivated these surgeons. First, was it possible to use local anaesthetic to perform a major operation? Would it block the pain sufficiently? It was difficult to find patients willing to volunteer to help answer this question, so it seemed only fair to the surgeons to use themselves as guinea pigs. Second, was major self-surgery even physically possible? The answer could be helpful for surgeons in isolated locations forced to save themselves. Each succeeding self-surgeon appeared to be unaware of his predecessors, so the same questions were answered again and again. Of course, a variety of less scientific motivations also drove the researchers – simple curiosity, adventure seeking, and showing off – so even if they had known of their predecessors, they might have proceeded anyway.

The trailblazer was Alexandre Fzaïcou, a surgeon at St Mary's Hospital in the Romanian city of Iaşi. On 29 September 1909, a colleague, Professor Juvara, gave him a spinal injection of anaesthesia and then Fzaïcou picked up his knife and went to work on himself to repair a hernia. His specific goal, he said, was to prove that consciousness, intelligence and manual dexterity were in no way diminished by an intra-spinal injection. The operation was a success. He suffered afterwards only from a slight headache that he attributed to the anaesthetic.

Three years later, Jules Regnault, chief of surgery at St Mandrier's Hospital in Toulon, performed a similar operation on himself, repairing a hernia, but instead of a spinal anaesthetic he used a subcutaneous injection of cocaine and morphine to block the pain. He also added a dramatic flourish. He gave a blow-by-blow narration of the surgery for the benefit of several colleagues who were standing by observing, and when he arrived at the thick nerve fibres attached to the inguinal ligaments, he paused the operation, grinned at his colleagues, and began strumming his fingers across the fibers as if he was playing the mandolin. He said he felt absolutely no pain as he played on his nerve-fibre instrument. The operation was a success, and after a few weeks recuperating in Normandy, Regnault fully recovered.

In the same year, Bertram Alden, chief surgeon at San Francisco's French Hospital, performed the first auto-appendectomy. Like Fzaïcou, he used spinal anaesthesia, claiming that he wanted to prove that it kept the mind clear and motor skills unaffected. However, he faced significant resistance from his colleagues, one of whom, Dr Mardis, became increasingly agitated as the surgery proceeded. Eventually Mardis threatened to leave the room if Alden didn't stop. Reluctantly, Alden handed his scalpel over to the other surgeon, who finished the job, denying Alden the pleasure of seeing the experiment through to the end.

The Remarkable Dr Kane

There was a nine-year hiatus in the self-surgery movement, but it returned forcefully to the headlines in 1921 thanks to the efforts of Evan O'Neill Kane, a name that would tower above all others in the annals of professional self-surgery due to his being, to this day, the only serial practitioner of the art. Kane was a member of the prominent Kane family of Pennsylvania. He served as chief surgeon at the Kane Summit Hospital, established in 1892 by his mother, in the town of Kane, Pennsylvania, founded in 1863 by his father. So he was local royalty – not someone any of the staff were going to disobey. His younger brother Thomas also worked at the hospital as a surgeon.

Kane got his first taste of self-surgery in 1919, when he was fifty-eight years old. His little finger grew infected and he decided to amputate it himself. However, his international fame as a self-surgeon came two years later when he developed chronic appendicitis. Initially he arranged for his brother to remove the inflamed organ, assisted by the hospital's top staff, but as he lay on the operating table, he had a sudden change of heart. As Kane put it, 'I announced to their consternation my determination to begin work on my own account and went ahead.'

To get a good view of his abdomen, Kane propped himself up with pillows and ordered the anaesthetist to push his head forward. He injected himself with cocaine and adrenalin, and then he swiftly cut through the superficial tissue, found the swollen appendix, and excised it. He allowed his staff to stitch up the wound, as he typically left this task to them anyway. The entire procedure took thirty minutes. Kane noted he probably could have done it more quickly if it weren't for the air of chaos in the operating room, as people milled around, unsure of what they were supposed to do.

There was only one slight moment of panic during the operation. It was caused by, as Kane put it, 'the unexpected protrusion of two or three knuckles of gut in the middle of my work'. This happened when he momentarily drew himself up into a sitting position. He had an assistant push the guts back into his abdomen and resumed the operation. Kane insisted the operation would have gone more smoothly if the other doctors, particularly his brother, hadn't been 'a little unduly excited', which caused him to feel rushed.

Kane made a swift recovery. Fourteen days later he was back in the hospital operating on other patients. The operation made headlines throughout the world, attracting far more attention than any previous self-surgery. Why this was the case isn't clear. Perhaps it was because of the fame of his family. But even without his embrace of self-surgery, Kane would have made a name for himself as a medical maverick. For instance, during the early days of flight, he conducted a self-experiment by flying to 5,000 feet to test his heart action at high altitude. He was one of the first to promote the use of radium to treat cancer. He invented a device to improve the administration of intravenous fluid infusions in emergency conditions, and he introduced the use of phonographs in the operating room, arguing that music would both soothe patients who were conscious during their surgery and save doctors from having to converse with them.

These were his better-received ideas, but he had a number of duds as well. He tried to popularize the use of asbestos bandages as surgical dressings, arguing that, since they were fireproof, they could be easily sterilized. Following his auto-appendectomy, he proposed that a law be passed requiring all children to have their appendices removed, whether or not they were ill. And in 1925, he began the practice of signing all his operations with a small India-ink tattoo scratched into the skin of his patients. His signature was the Morse code symbol for the letter K (dash dot dash). He suggested surgeons should adopt a standardized tattoo code, which they could use to write pertinent medical facts on the skin of patients. Such readily viewable information would certainly be useful in the event of emergency operations, but patients resisted having their medical histories inscribed on their bodies, so the idea never caught on.

In 1932, when Kane was seventy-one years old, he realized he needed an operation to repair a hernia caused by a horse-riding accident. Despite his advanced age, and the fact that a herniotomy is a far more dangerous operation than an appendectomy because of the risk of cutting the femoral artery, it didn't surprise anyone when he decided to do it himself. At least this time he warned everyone before he was on the operating table. So he became the first and only person ever to self-operate on both an appendix and a hernia.

During the operation, Dr Howard Cleveland stood by, ready to take over in case anything went wrong. Also present were Kane's son Philip, several nurses, and a reporter. Kane propped himself up on the operating table in a half-sitting position and began the procedure without hesitation, joking with nurses as he sliced through his tissue. He showed pain only once, when the anaesthetic started to wear off, and he called for another shot. When he reached a critical stage of the operation, he dramatically announced, 'The risk is here and I must face it.' The operation ended without event, after an hour and forty-five minutes. As before, Kane didn't close the wound himself, allowing Dr Cleveland to do this. Cleveland also added the final flourish by tattooing Kane's signature mark above the incision.

Initially Kane seemed to recover well from the operation. He was back on his feet two days later, assisting Dr Cleveland with an operation. However, he never fully regained his health. Less than three months later, still in a weakened condition, he came down with pneumonia and died.

The End of the Auto-Surgery Fad

Although Dr Kane's successful auto-appendectomy in 1921 received international media coverage, it didn't inspire many other surgeons to imitate him. In fact, only two other doctors appear to have followed in his footsteps. In January 1929, Dr David Rabello, chief surgeon at São Vicente de Paulo Hospital in Belo Horizonte, Brazil, showed up in the operating room at 8 a.m. and told the interns, students, and nurses gathered there that they were going to have the chance to witness an interesting hernia operation, and that he would be the patient. 'Who is performing the operation?' they asked. 'The same Rabello,' he jauntily replied. The operation was a success.

Almost twenty years later, in December 1948, Dr Theodor Herr of Hamburg removed his own appendix in a near five-hour operation. Like Kane, he hunched over in a half-sitting position as he worked. He told the press he operated on himself purely 'out of professional curiosity'. It's not clear why he required so much time to close himself up.

So what brought the brief enthusiasm for experimental self-surgery to a close? Probably the novelty simply wore off. It was, after all, little more than a stunt from which surgeons had nothing to gain, and much to lose. (It's likely some surgeons still perform minor self-surgeries on themselves, but this activity doesn't get reported in medical journals.) However, it's curious, though not necessarily meaningful, that the end of the self-surgery movement coincided with the end of the widespread use of cocaine in operating rooms. After the 1940s, surgeons favoured safer local anaesthetics (that didn't run the risk of addiction) such as novocaine and lidocaine. Without cocaine, self-surgery perhaps just didn't seem as exciting as it once had.

For whatever reason, by the 1950s, self-surgery once again was practised exclusively by the desperate and insane. Among the desperate was a young Soviet surgeon, Leonid Ivanovich Rogozov, who, in April 1961, developed an inflamed appendix while stationed at the Novolazarevskaya Research Station on the Queen Astrid coast of Antarctica. Unable to return to the main base because of a snowstorm, he faced the reality that if he didn't remove his appendix himself he was going to die. He might have been comforted had he known that other surgeons had successfully removed their own appendices in the past, but he wasn't aware of this. As far as he knew, he was a reluctant medical pioneer. But unlike his predecessors, Rogozov had no professional help standing by. His assistants were a terrified meteorologist and a mechanic. So as the Antarctic wind howled outside, the young surgeon nervously cut himself open. Thankfully everything went well. He returned to the Soviet Union a national hero.

A similar case occurred in 1999 when Dr Jerri Nielsen, stationed at the Amundsen-Scott South Pole Station, discovered a lump in her right breast. Doctors back in the United States, with whom she consulted via satellite link, advised her the lump should be biopsied. However, she was the only doctor at the station, and the Antarctic winter prevented planes from landing for almost half a year, which meant she'd have to operate on herself, which she proceeded to do. Subsequently, upon learning the tumor was cancerous, she self-administered chemotherapy. A ski-equipped plane, sent weeks ahead of schedule, finally brought her home to safety, and to a new career as a motivational speaker.

As long as we keep sending surgeons into remote, inaccessible locations – the Antarctic, or perhaps long-term space missions to the moon or Mars – it's likely more doctors will eventually have to turn the knife upon themselves. However, it seems unlikely that experimental self-surgery will ever experience a revival. Though who knows. There may come a day when advances in robotic, remote-controlled surgery make it possible for a surgeon to do something previously unimaginable, such as operating on his own brain. The temptation to perform auto-brain surgery might well prove irresistible to some future daredevil surgeon.

Killing Yourself for Science

Omaha, Nebraska – 25 November 1936. Edwin Katskee sits in his office. He's a young man in his early thirties with elegant features and dark, wavy hair. For several minutes he stares blankly at the framed degree certificates hanging on his wall and the rows of medical volumes on his bookshelf. Then he looks down at the syringe and vial lying in front of him on the desktop and sighs.

He picks up the phone and dials the number for the building's night attendant. When the attendant picks up the line, he says, 'Hello, John. It's Dr Katskee. I wanted to thank you for coming up and assisting me with my project earlier.' He pauses to let the other man reply, and then he resumes speaking.

'Well, I also wanted to let you know that I'll be staying in my office a while longer. I'm doing some research. Oh, and I may write some notes on the wall. Please don't let anyone remove them after I leave. They're quite important.' The tinny voice of the attendant speaks into his ear, offering assurances of cooperation.

'Anyway, that's all I wanted to say . . . Well, I hope you have a happy Thanksgiving tomorrow too, John. Give my regards to your wife and family . . . No, I probably won't be doing anything. I'm quite busy with my work . . . Yes, that's all. Thank you again, John.'

Katskee places the phone back on the receiver. For several minutes he doesn't move, lost in thought. At last he nods his head, as if affirming a decision he's made, stands up, and looks at his watch. He picks up a pen, walks over to the wall of his office, and begins to write on the blank vertical surface. His handwriting is bold and firm.

'10 p.m. I do this for a better understanding of the bad reactions we see in rectal patients when cocaine is applied topically or injected, when novocaine is ineffective.'

He steps back to consider what he's written, then leans forward and adds another line.

'This is just my way of contributing to medical and scientific archives of clinical research.' Finally he signs his name, 'Katskee.'

Satisfied with this, he walks back to the desk, lifts up the syringe, and fills it from the vial. Then he rolls up his shirtsleeve and plunges the needle into the flesh of his arm. He clenches his teeth as he feels the drug enter his body. As soon as the syringe is empty, he places it back down on the desk, picks up his pen, returns to the wall, and again starts writing.

'Vision clear – slight ringing in ears – heartbeat increasing – urge to talk.'

He steps back and bumps into the desk. Nervously he looks around, sees nothing, and then he begins to pace back and forth across the carpet, muttering to himself as he does so. 'Thoughts focused . . . Slight numbness . . .'

He extends his arms back and forth, like someone doing stretching exercises. More cryptic phrases emerge from his mouth. 'They better pay attention . . . Got the instructions for the pulmotor . . . Make sure they see them . . .'

Abruptly he rushes back to the wall and writes more notes. 'Ringing getting louder . . . Loquacious . . . Unusual . . . Normally quiet as a person under ether . . .'

Suddenly he stops. His face contorts with pain. His legs buckle, and he falls to his knees.

'Oh, God,' he whispers. 'Need to call John.'

He continues to move his mouth, trying to speak, but no words come out. He grips his neck with his left hand. His right hand trembles. The tremors move up his arm towards his shoulder. He twitches once and collapses to the ground, his body rigid. His face is bright red. His mouth gapes open, struggling for air. His eyes dart back and forth, searching for help. A long, rattling gasp escapes from his throat. Then his entire body begins to shake, convulsed by violent, uncontrollable spasms.

Science can be a dangerous profession. Lethal bacteria, radioactive materials, dangerous animals, and toxic chemicals are just a few of the things a researcher can run foul of. One slip, one careless accident, and the results can be disastrous – even fatal. A book could easily be filled with cases of men and women dead as a result of their scientific research.

Most researchers try to minimize the risks. They use every precaution available. They wear masks and gloves, breathe through filters, hide behind bio-shields, or deploy remote-controlled equipment. Of course, there have been cases in which researchers have shrugged off such safeguards and taken enormous, even foolish, risks. In order to prove a theory, researchers have swallowed vials of dangerous bacteria or exposed themselves to radiation. Often they get lucky, but on occasion they've miscalculated the risks and paid the ultimate price. But even in such cases the researchers didn't actively want to die. They weren't trying to kill themselves. However, there are a few cases in which that's exactly what they wanted. Sometimes the experience of death – what it feels like to have life ebb away – becomes the focus of a scientist's curiosity, and when that's the case the only way he (it seems invariably to be men involved in this pursuit) can satisfy his interest is to kill himself, or at least come very close to doing so.

Death by Hanging

In 1623, Sir Francis Bacon described an unusual experiment conducted by an unnamed gentleman who wanted to experience the sensation of hanging. This gentleman was curious about what it would feel like as the noose tightened around his neck and death crept up on him. However, he didn't want to die, so he convinced a friend to help him satisfy his morbid inquisitiveness.

As his friend stood by, ready to come to his aid, the self-experimenter wrapped a rope around his neck, tied the other end to a beam overhead, and stepped off a stool. He thought it would be a simple matter to regain the safety of the stool, but instead his arms and legs flailed uncontrollably. Within seconds he blacked out. His friend rushed to his aid, cut the rope, and helped him to the ground. Later, after the gentleman returned to consciousness, he said that he hadn't experienced any pain, but that it had seemed at first as if fire was burning all around him, then the fire was replaced by 'an intense blackness or darkness, and then by a kind of pale blue or sea-green colour, such as is often seen also by fainting persons'.

Hanging, unlike other methods of self-harm – bleeding, electrocution, poisoning, etc. – allows a measure of control. As long as proper precautions are in place, the researcher can stop the experiment before doing irreversible damage to himself. For this reason, hanging has proven to be a popular method used by scientific explorers of the near-death experience.

In 1832, a French expert in forensic medicine, Monsieur Fleichmann, repeated the experiment of Sir Francis's anonymous gentleman. Fleichmann was conducting a study of 'different kinds of death by strangulation', and he reasoned it would be helpful if he knew what hanging felt like. So he placed a cord around his neck, directly beneath his chin, and leaned upon it just enough to cut off his blood circulation but not his breath. Almost at once, a sense of enormous weight pressed down on his lower extremities and heat suffused his head. He heard a loud ringing sound in his ears. His face turned scarlet, and his eyes bulged out. He claimed to have endured two minutes of this before his self-preservation instinct kicked in, and he raised his neck off the rope.

When the New York City doctor Graeme Hammond arranged for several of his friends to strangulate him in 1882, he explained that he hoped his experiment would produce information that would make the execution of criminals more swift and humane. Specifically, he wanted to determine whether compression of the blood vessels and windpipe alone provides a quick, painless death, or whether it's necessary for the hangman to also dislocate the neck of the victim. Hammond was twenty-four years old, a year out of medical school, and his experiment had the feel of youthful bravado.

Hammond sat in a chair while a friend wrapped a towel around his neck and twisted the ends together to pull it increasingly tight. A second friend stood in front of him to monitor his facial expressions, while simultaneously stabbing him repeatedly in the hand with a knife to measure his sensitivity to pain. As the towel pinched into his neck, Hammond experienced a warm, tingling sensation that began in his feet and rapidly spread throughout his body. His vision grew clouded. There was a roaring in his ears, and his head felt like it was about to burst. After one minute and twenty seconds 'all sensibility was abolished'. His friends stopped to allow him a few minutes of rest. Then they repeated the test. This time Hammond lost all sensation after only fifty-five seconds. His friend stabbed his hand hard enough to draw blood, but Hammond felt nothing. When the ordeal ended, Hammond concluded, 'In order to obtain a speedy and painless death it is neither necessary nor desirable that the neck should be dislocated.'

Hammond went on to have a distinguished career as the physician to several US Olympic teams, but another near-death experience brought his name back into the newspapers. In December 1913, he stepped into an elevator at the New York Athletic Club. Suddenly a cable snapped, and the elevator plunged down three storeys. Luckily, it got wedged in the shaft before it hit the bottom, and Hammond walked away unharmed.

The experience seemed to convince him he was death-proof, because he subsequently began to boast about the extraordinary powers of his body, claiming he was able to shrug off the effects of harmful substances. 'I do all the things that are bad for me. . . . I go to banquets and eat and drink my fill. But my constant exercise comes to the rescue. Through it my system throws off all the things that are bad for me.' At seventy-seven, still going strong, his self-assessment appeared justified. He told the press, 'On my birthdays I do four miles just to prove to myself that I am not getting old. And I intend to keep it up as long as I live, which probably won't be much over 110.' At eighty he was still a regular in the gym. But in 1944, at the age of eighty-six, death finally caught up with him. He slipped into a coma and died just days after the death of his beloved daughter.

As extreme as the strangulation experiments of Fleichmann and Hammond were, they paled in comparison to the feats of the Romanian doctor, Nicolae Minovici. In the first decade of the twentieth century, while employed as a professor of forensic science at the State School of Science in Bucharest, Minovici undertook a comprehensive study of death by hanging. Like his predecessors, he felt compelled to experience for himself what he was researching, as if dancing right up to the edge of death was too great a temptation to resist.

He constructed a system that allowed him to auto-asphyxiate – a hangman's knot tied in the end of a rope that ran through a pulley attached to the ceiling. He lay down on a cot, placed his head through the noose, and firmly tugged the other end of the rope, pulling hard enough to lift his head and shoulders off the cot. The noose tightened around his neck, his face turned a purple-red, his vision blurred, and he heard a whistling. He lasted only six seconds doing this before consciousness began to slip away, forcing him to stop.

However, this wasn't enough. He wanted the full, swinging-by-his-neck experience. So he arranged for assistants to help him achieve it. He tied a non-constricting knot in the rope, again placed his neck in the noose, and gave the signal to begin. The assistants pulled with all their might, and he rose by his neck two metres off the ground. He later described the sensation:

> As soon as my feet lifted off the ground, my eyelids shut violently, and my respiratory tract closed so hermetically that it was impossible to breathe. I could not even hear the voice of the employee holding the rope who was counting the seconds out loud. My ears were whistling and both the pain and the need to breathe led me to cut the experiment short because I could not bear it anymore. I had to get down.

On this first occasion, he lasted only a few seconds before his courage failed and he gave the signal to stop, but it became a personal challenge for him to last longer. He noted that Fleichmann had asphyxiated himself for a full two minutes, and this feat taunted him. Minovici, evidently competitive by nature, reasoned that he should easily be as tough as that. So day after day, he repeated the experiment, seeking to beat his record. Finally, after twelve days, he endured a full twenty-five seconds in the air. Realizing he had reached his limit, he concluded that Fleichmann must have lied. He wrote, 'While I am in agreement with Fleichmann as to what the symptoms of hanging are (which he has been able to witness on himself as have I), I cannot admit that there is a possibility one could go through this experiment for two minutes without losing consciousness at a much earlier point.'

However, Minovici wasn't quite finished. One final experiment tempted him, the ultimate test – hanging from the ceiling by a constricting hangman's knot. He tied the knot, again placed his head through the noose, and gave his assistants the signal. They pulled. Instantly a burning pain ripped through his neck. The constriction was so intense that he frantically waved his assistants to stop. He had only endured four seconds. His feet hadn't even left the ground. Nevertheless, the trauma to his neck made it painful for him to swallow for an entire month.

Minovici published the results of his study in 1905, in a magisterial 218-page tome, *Étude Sur la Pendaison* (*A Study of Hanging*). It instantly became the classic reference work on the subject – not that it faced much competition. The book is available in its entirety online and is worth seeking out, even if you don't read French, because of the photographs of his self-experimentation that Minovici included. He was a handsome man, with a thick head of hair and a full, bushy moustache, and we see him strangling himself while lying on a cot, swinging by his neck from the ceiling, and then proudly showing off the prominent bruise around his neck. He did all this while dressed in expensive suit pants and an elegant waistcoat, having apparently only removed his jacket and tie for the sake of the experiment. It's evident he didn't want readers to mistake him for a member of the lower classes, and in fact Minovici was best known in Romania as a wealthy patron of the arts. With his personal fortune he founded a Museum of Folk Art, which remains in existence today.

Death by Freezing

The convenience and reversibility of hanging may have made it the preferred method of investigating the near-death experience, but of course there are other ways of approaching the margins of death, and researchers have dutifully explored these. Most notably, the Cambridge physiologist Sir Joseph Barcroft acquired a reputation as a specialist in death-defying experiments. He called these his 'borderline excursions'.

Barcroft's first experimental brush with death occurred during World War I, when he volunteered to be exposed to hydrocyanic acid gas (aka prussic acid). A dog in the gas chamber with him died in ninety-five seconds, but Barcroft waited ten minutes before stumbling out with the dog in his arms, announcing he felt giddy. In 1920, he sealed himself inside an airtight glass cage to test the minimum amount of oxygen needed to survive. His colleagues thinned the air to the equivalent of an altitude of 16,000 feet, and Barcroft endured this environment for six days, until his entire body turned a deep shade of blue. Later that same decade, he sat for twenty minutes inside a chamber containing an atmosphere of 7.2 per cent carbon dioxide, which caused him to develop a severe headache and feel disoriented. He said he had no desire to repeat that test. However, his closest brush with death occurred in the 1930s when he decided to freeze himself.

Barcroft stripped naked and lay down on a couch in a freezing room with the windows open. At first he shivered and curled up to stay warm. He found it difficult to maintain the willpower to remain in the room. He kept thinking, 'I could just walk out of here now,' but he persevered, and after about an hour a strange mental change occurred. In 1936, during a lecture at Yale University, he described what happened:

> As I lay naked in the cold room I had been shivering and my limbs had been flexed in a sort of effort to huddle up, and I had been very conscious of the cold. Then a moment came when I stretched out my legs; the sense of coldness passed away, and it was succeeded by a beautiful feeling of warmth; the word 'bask' most fitly describes my condition: I was basking in the cold.

The sensation he felt indicates he was quite close to death. It's a phenomenon often seen in cases of lethal hypothermia. Shortly before death, as the nerve endings go haywire, a person feels intense heat, as if burning up. Victims of freezing often rip off all their clothes and crawl around in the snow and ice in an effort to cool down. Physiologists call the behaviour 'paradoxical undressing'.

Thankfully for Barcroft, one of his assistants noticed something was amiss and rushed in with a blanket and warm drink to save him. Barcroft lived on to the age of seventy-four, finally collapsing on a bus while riding home from work.

Death by Fever

The researchers discussed so far all inched up to the very edge of death, but they stopped at the border, having no desire to proceed further. They wanted to taste death, but not accept its embrace completely. However, there are a few sad cases of scientists who attempted to press onward, purposefully blurring the line between self-experimentation and suicide.

Élie Metchnikoff was a Russian biologist who conducted pioneering work on the immune system. In 1881 his wife fell ill with typhoid fever, and while Metchnikoff nursed her back to health, he fell into a state of severe depression, during which he decided to take his life. The method he settled on was to infect himself with relapsing fever. He figured this would kill two birds with one stone, so to speak. It would disguise his suicide, thus sparing his family that agony, and it would also settle the issue of whether relapsing fever could be transmitted through the blood. So he infected himself and waited to see what would happen.

The answer was that, yes, relapsing fever can be transmitted through the blood. Metchnikoff became deathly ill, and in his delirium he experienced a strange revelation, later described by his wife (who had recovered from her illness) in her biography of him: 'He had a very distinct prevision of approaching death. This semi-conscious state was accompanied by a feeling of great happiness; he imagined that he had solved all human ethical questions.

Much later, this fact led him to suppose that death could actually be attended by agreeable sensations.'

Luckily for Metchnikoff, his experiment didn't fully succeed because his body fought off the illness. He lived on to win the Nobel Prize in Medicine in 1908 for his work on immunity, and he died of natural causes in 1916.

Death by Cocaine

The Nebraska proctologist Edwin Katskee wasn't as lucky as Metchnikoff. Katskee was a young doctor with a promising career ahead of him. At the age of thirty-four, he was a frequent contributor to medical journals, including among his publications an article in the *American Journal of Surgery* titled, 'Anorectal Fistulectomy: A New Method'. The only sign of trouble in his life was the recent breakup of his marriage. Then, on the night of 25 November 1936, Katskee decided to conduct a risky self-experiment. He injected himself with a potentially lethal amount of cocaine.

The last person to see Katskee alive was the night attendant in the office building where he worked. Responding to a call from Katskee, the attendant came up and helped take the young doctor's blood pressure and test his reflexes. It was a slightly unusual request, but the attendant thought nothing of it until Katskee was found dead in his office the next day. But what caught the media's attention was that Katskee's actions didn't appear to be a simple, unambiguous case of suicide, because on an empty wall of his office Katskee left behind a detailed clinical account of his thoughts and sensations as he advanced towards death. The newspapers called it his 'death diary'.

Katskee scrawled the notes in no apparent order, but it was possible to piece together their chronology by the decreasing legibility of his handwriting as the end approached. In an early note he stated clearly that he intended his actions to be a form of scientific experimentation, explaining that he hoped to contribute to a better understanding of the occasional bad reactions seen in patients when cocaine was applied topically or injected. Katskee also left instructions for those who might find him: 'Have a university or any med college pharmacologist give you an opinion on my findings. They better be good because I'm not going to repeat the experiment.'

As the drug took effect, he recorded effects such as: 'Eyes mildly dilated. Vision excellent.' The drug caused bouts of paralysis and convulsions that came in waves. In between one of these bouts he wrote, 'Partial recovery. Smoked cigarette.' High up on the wall he scribbled, 'Now able to stand up.' And elsewhere, 'After depression is terrible. Advise all inquisitive M.D.'s [sic] to lay off this stuff.' In one spot, in a shaky hand, he recorded his 'Clinical course over about twelve minutes':

Symptoms: Convulsions followed by paralysis of tongue.
(1) Speech and only tongue movable. Can't understand that movable tongue and no speech. Voice O.K.
(2) Staggering gait preceding paralysis.
(3) Paralysis . . .

The word 'paralysis' tapered off into a wavy scrawl that descended to the floor. It was, perhaps, the last word he ever wrote.

Did Katskee intend to commit suicide? He clearly realized people would wonder about this because in one of his notes he insisted, 'Narcotics poisoning – not suicide.' In a notebook on his desk, he had also left instructions detailing how he could be revived. His family refused to believe he had taken his life, calling him a 'martyr to science'. But a medical colleague pointed out that Katskee must have known that the amount of cocaine he injected would inevitably prove fatal.

If Katskee hadn't intended to kill himself, if he sincerely believed his experiment to be a contribution to science, then his death was even more tragic, because when Dr Charles Poynter of the University of Nebraska College of Medicine examined the notes left on the wall, he concluded they were so incoherent as to be of no value whatsoever. Another colleague, Dr Herman Jahr, gave a slightly more positive appraisal, declaring the notes 'might be interesting', but Jahr was probably trying to be kind.

Death by Physics

As strange as the death of Katskee was, it isn't the strangest example of a suicide experiment. An even odder example might be found in an unlikely place – the abstract, mathematical realm of quantum physics – but only *might be found* because the case is so peculiar that it's not clear whether it was a suicide at all.

To understand the case, some familiarity with the concepts of quantum physics is necessary. During the early twentieth century, physicists observed a puzzling paradox. Light photons, depending on the method of experimentally observing them, sometimes moved like a particle (in a straight line) and at other times like a wave (in an oscillating pattern). This didn't make sense to physicists, because how could anything be both a wave and a particle? During the 1920s, a group of researchers in Copenhagen offered a possible explanation. Light photons, they argued, occupy a variety of contradictory states (both particle and wave) simultaneously. They exist in a state of 'quantum superposition'. Only when observed do they collapse into a single state. The act of observation, the Copenhagen scientists suggested, actually causes the photon to become either a particle or a wave.

In 1935, the Austrian physicist Erwin Schrödinger devised a thought experiment to illustrate how counterintuitive and bizarre the concept of quantum superposition was. Imagine, he said, a cat sealed inside a box with a small amount of radioactive material, a Geiger counter, and a vial of poisonous gas. Within the course of an hour, there's a 50 per cent chance one of the atoms in the radioactive material will decay. If it does so, the Geiger counter will detect the decay and activate a mechanism to smash the vial of poisonous gas, thereby killing the cat. However, there's an equal probability nothing will happen and the cat will remain alive. So at the end of an hour, is the cat alive or dead? Common sense suggests it must be one or the other. According to quantum physics, however, the cat is both alive and dead at the same time. It's in a state of quantum superposition until someone lifts the lid of the box and observes its condition.

Schrödinger intended his thought experiment to be a purely hypothetical example. He certainly didn't intend that anyone should conduct the experiment, let alone substitute a human being for the cat. But it's possible that three years later, in 1938, a brilliant young Sicilian physicist, Ettore Majorana, did essentially that, transforming himself into the real-world equivalent of Schrödinger's cat.

As noted, Majorana was brilliant – he was, for instance, the first researcher to predict that neutrinos have mass – but he was also highly eccentric. He hated accepting credit for his work, though his work was of Nobel Prize calibre, and he was reclusive to the point of living like a hermit, completely shunning the company of all other people. Throughout the 1930s, his withdrawal from society deepened. Then, on 25 March 1938, at the age of thirty-one, Majorana boarded a ferry from Palermo to Naples. He never disembarked. Somehow he disappeared during the trip, never to be seen again.

Majorana's friends and family, once they realized he was gone, frantically tried to figure out what had happened to him. The only clue was a mysterious series of letters and telegrams he had written shortly before his departure and sent to Antonio Carrelli, director of the Naples Institute of Physics where he taught. In the first letter, Majorana announced he had made an 'inevitable' decision to disappear and apologized for the inconvenience he realized this would cause. But soon after this, apparently having changed his mind, Majorana cabled a telegram to the director, urging him to ignore the previous letter. Then he sent a final letter: 'Dear Carrelli, I hope you got my telegram and my letter at the same time. The sea rejected me and I'll be back tomorrow at the Hotel Bologna travelling perhaps with this letter.'

Majorana's friends and family didn't know what to make of

these letters, but as the days turned into weeks, and then into months, they reluctantly concluded that the young physicist had committed suicide by jumping overboard while en route to Naples. No one witnessed him jump, but how else to explain the reference to the sea in his final letter?

As the years passed, Majorana became like the Elvis Presley of the physics world. Increasingly speculative theories circulated about his possible fate. Some people suggested Nazi agents had killed him, to prevent him from helping the Allies build an atomic bomb. Others argued that he joined a monastery, or fled to Argentina to start a new life. There were even reported sightings of a mysterious beggar in Naples who helped local students with their mathematics. It was only recently, in 2006, that the theoretical physicist Oleg Zaslavskii pointed out the intriguing similarities between Majorana's disappearance and Schrödinger's cat-in-a-box thought experiment – a thought experiment Majorana certainly knew about. In fact, in 1938, he would have been one of the few people in the world to fully understand its implications.

Zaslavskii noted that the circumstances of Majorana's disappearance left investigators only two possibilities. Either Majorana committed suicide by jumping overboard, or he disembarked in Naples and went into hiding. Put another way, these choices eerily match the two possible states of a light photon. That is, Majorana either fell into the waves, or he walked off the boat in a straight line.

Such a similarity might simply be a coincidence, but what struck Zaslavskii as more peculiar were Majorana's last communications. First Majorana sent a letter saying he had decided to disappear. He followed this with a telegram announcing a change of mind, and finally he sent a letter expressing the hope that Carrelli had learned of the two possibilities *at the same time.* Wouldn't it have made more sense to hope that the telegram renouncing his intention to disappear had arrived first? Why would Majorana want Carrelli to learn of the two possibilities together? The answer, Zaslavskii suggested, was that Majorana deliberately arranged events so that the director would have to consider two contradictory alternatives simultaneously. By doing so, Majorana placed himself in a state of quantum superposition, just like Schrödinger's cat.

At first blush, Zaslavskii's theory sounds far-fetched, but is it really? If Majorana was eccentric enough to engineer his own disappearance, he certainly could have planned to do so with a parting rhetorical flourish for the benefit of his fellow physicists, vanishing paradoxically, like a cat inside a box. After all, physics was his entire life – and apparently his death too. And whether Majorana purposefully intended to mimic Schrödinger's cat or not, the fact is, that's what he achieved. On the night of 25 March, he set sail, and by the end of his journey he had ceased to be any one thing, but instead became several different things, both alive and dead, simultaneously.

SELECTED REFERENCES

One: Electric Bodies

ELECTRIFYING BIRDS

Cheney, M. (1981), *Tesla: Man Out of Time*, Englewood Cliffs, NJ: Prentice-Hall, Inc.

Driscol, T. E., O. D. Ratnoff, & O. F. Nygaard (1975), 'The Remarkable Dr. Abildgaard and Countershock', *Annals of Internal Medicine*, 83: 878–82.

Elsenaar, A., & R. Scha (2002), 'Electric Body Manipulation as Performance Art: A Historical Perspective', *Leonardo Music Journal*, 12: 17–28.

Franklin, B., *The Papers of Benjamin Franklin*. Available online at http://www.franklinpapers.org.

Gray, S. (1735), 'Experiments and Observations upon the Light that is Produced by Communicating Electrical Attraction to Animal or Inanimate Bodies, Together with Some of its Most Surprising Effects', *Philosophical Transactions*, 39: 16–24.

Heilbron, J. T. (1982), *Elements of Early Modern Physics*. Berkeley: University of California Press.

Jex-Blake, A. J. (1913), 'Death by Electric Currents and by Lightning', *The British Medical Journal*, 1(2724): 548–52.

Montoya Tena, G., R. Hernandez C., & J. I. Montoya T. (2010), 'Failures in Outdoor Insulation Caused by Bird Excrement', *Electric Power Systems Research*, 80: 716–22.

Needham, J. T. (1746), 'Concerning Some New Electrical Experiments Lately Made at Paris', *Philosophical Transactions*, 44: 247–63.

Riely, E. G. (2006), 'Benjamin Franklin and the American Turkey', *Gastronomica: The Journal of Food and Culture*, 6(4): 19–25.

Schiffer, M. B. (2003), *Draw the Lightning Down: Benjamin Franklin and Electrical Technology in the Age of Enlightenment*. Berkeley: University of California Press.

Watson, W. (1751), 'An Account of Mr. Benjamin Franklin's Treatise, Lately Published', *Philosophical Transactions*, 47: 202–11.

Winkler, J. H. (1746), 'Concerning the Effects of Electricity upon Himself and his Wife', *Philosophical Transactions*, 44: 211–12.

THE MAN WHO MARRIED HIS VOLTAIC PILE

Christensen, D. C. (1995), 'The Ørsted-Ritter Partnership and the Birth of Romantic Natural Philosophy', *Annals of Science*, 52(2): 153–85.

Deeney, N. (1983), 'The Romantic Science of J. W. Ritter', *The Maynooth Review*, 8: 43–59.

Ritter, J. W. (1802), *Beyträge zur nähern Kenntniss des Galvanismus und der Resultate seiner Untersuchung*, Vol. 2, Jena: Friedrich Frommann.

Strickland, S. W. (1992), *Circumscribing science: Johann Wilhelm Ritter and the Physics of Sidereal Man*, Cambridge, MA: Harvard University. Ph.D. dissertation.

——— (1998), 'The Ideology of Self-Knowledge and the Practice of Self-Experimentation', *Eighteenth-Century Studies*, 31(4): 453–71.

Trommsdorff, J. B. (1803), *Geschichte des Galvanismus*. Erfurt: Henningschen Buchhandlung.

Trumpler, M. J. (1992), *Questioning Nature: Experimental Investigations of Animal Electricity, 1791–1810*, New Haven: Yale University. Ph.D. dissertation.

Wetzels, W. D. (1973), *Johann Wilhelm Ritter: Physik im Wirkungsfeld der deutschen Romantik*. Berlin: Walter de Gruyter.

——— (1990), 'Johann Wilhelm Ritter: Romantic Physics in Germany', in Cunningham, A., & N. Jardine, eds., *Romanticism and the Sciences*, Cambridge: Cambridge University Press, 199–212.

ELECTROCUTING AN ELEPHANT

Brown, H. P. (1888), 'Death-current Experiments at the Edison Laboratory', *The Medico-Legal Journal*, 6: 386–9.

'Coney Elephant Killed' (5 January 1903), *New York Times*: 1.

'Coney Swept by $1,500,000 Fire' (29 July 1907), *New York Times*: 1.

'Electrifying Animals: How Monkeys and Elephants Act when under the Influence of a Battery' (15 March 1889), *Daily Independent* (Monroe, WI): 3.

'Elephant Skull Dug Up: Luna Park Herd Uneasy while it Remained Buried' (7 August 1905), *New York Tribune*: 5.

Essig, M. (2003), *Edison & the Electric Chair*, New York: Walker & Company.

'Ghost of Elephant Haunts Coney Island' (15 February 1904), *Evening News* (San Jose, CA): 2

McNichol, T. (2006), *AC/DC: The Savage Tale of the First Standards War*, San Francisco: Jossey-Bass.

Moffett, C. (23 June 1895), 'Elephant Keeping', *Los Angeles Times*: 26.

FROM ELECTRO-BOTANY TO ELECTRIC SCHOOLCHILDREN

'China to Send Pig Sperm to Space' (17 July 2005), BBC News. http://news.bbc.co.uk/2/hi/asia-pacific/4690651.stm

de la Peña, C. T. (2003), *The Body Electric: How Strange Machines Built the Modern American*, New York: New York University Press.

Demainbray, S. (1747), 'Of Experiments in Electricity', *The Gentleman's Magazine*, 17: 80–1, 102.

'Electricity for Children' (17 November 1912), *Washington Post*: M1.

'Electricity in Relation to Growth' (June 1918), *Current Opinion* 64(6): 409.

'Electricity Makes Chickens Grow Big' (18 February 1912), *New York Times*: C4.

'Electrified Chickens: Electricity as a Growth Stimulator' (11 October 1913), *Scientific American* 109(15): 287.

'Electrified Schoolroom to Brighten Dull Pupils' (18 August 1912), *New York Times*: SM1.

'Novel Application of Electricity' (1869), *New England Medical Gazette*, 4(3): 102–3.

Spence, C. C. (1962), 'Early Uses of Electricity in American Agriculture', *Technology and Culture*, 3(2): 142–60.

LIGHTNING, CHURCHES, AND ELECTRIFIED SHEEP

Andrews, C. (1995), 'Structural Changes after Lightning Strike, with Special Emphasis on Special Sense Orifices as Portals of Entry', *Seminars in Neurology*, 15(3): 296–303.

'Fair Appreciated by Helen Keller' (1 November 1939), *New York Times*: 17.

Friedman, J. S. (2008), *Out of the Blue: A History of Lightning – Science, Superstition, and Amazing Stories of Survival*, New York: Bantam Dell.

Hackmann, W. D. (1989), 'Scientific Instruments: Models of Brass and Aids to Discovery', in Gooding, D., T. Pinch, & S. Schaffer, eds., *The Uses of Experiment: Studies in the Natural Sciences*, Cambridge: Cambridge University Press, 31–66.

Henley, W. (1774), 'Experiments Concerning the Different Efficacy of Pointed and Blunted Rods, in Securing Buildings Against the Stroke of Lightning', *Philosophical Transactions* 64: 133–52.

Howard, J. R. (1966), *The Effects of Lightning and Simulated Lightning on Tissues of Animals*. Ames, IA: Iowa State University of Science and Technology. Ph.D. dissertation.

Langworthy, O. R., & W. B. Kouwenhoven (October 1931), 'Injuries Produced in the Organism by the Discharge from an Impulse Generator', *Journal of Industrial Hygiene*, 13(8): 326–30.

Lavine, S. A. (1955), *Steinmetz: Maker of Lightning*, New York: Dodd, Mead & Company.

'Lightning Is this Wizard's Plaything' (10 June 1923), *New York Times*: XX3.

'Modern Jove Hurls Lightning at Will' (3 March 1922), *New York Times*: 1.

Read, J. (1793), *A Summary View of the Spontaneous Electricity of the Earth and Atmosphere*, London: The Royal Society of London. 71–2.

Scherrer, S. J. (1984), 'Signs and Wonders in the Imperial Cult: A New Look at a Roman Religious Institution in the Light of Rev 13:13–15', *Journal of Biblical Literature*, 103(4): 599–610.

Schiffer, M. B. (2003), *Draw the Lightning Down: Benjamin Franklin and Electrical Technology in the Age of Enlightenment*, Berkeley: University of California Press.

Wall, W. (1708), 'Experiments of the Luminous Qualities of Amber, Diamonds, and Gum Lac', *Philosophical Transactions*, 26: 69–76.

Warner, D. J. (August 1997), 'Lightning Rods and Thunder Houses', *Rittenhouse*, 11(44): 124–7.

Two: Nuclear Reactions

PSYCHONEUROTIC ATOMIC GOATS

'Artificial Blitzkriegs' (8 September 1941), *Reno Evening Gazette*: 4.

Gantt, W. H. (1944), *Experimental Basis for Neurotic Behavior*, New York: Paul B. Hoeber, Inc.

Gerstell, R. (7 January 1950), 'How You Can Survive an A-Bomb Blast', *Saturday Evening Post*, 222(28): 23, 73–5.

'Hate Training Has Ceased' (9 July 1942), *Ottawa Citizen*: 22.

'How to Have a Breakdown' (19 February 1950), *Parade*: 10–11.

Liddell, H. S. (1944), 'Conditioned Reflex Method and Experimental Neurosis', in Hunt, J. M., ed., *Personality and the Behavior Disorders*, Vol. 1, New York: The Ronald Press Company, 389–412.

——— (1960), 'Experimental Neuroses in Animals', in Tanner, J. M., ed., *Stress and Psychiatric Disorder*, Oxford: Blackwell Scientific Publications, 59–64.

McLaughlin, F. L., & W. M. Millar (2 August 1941), 'Employment of Air-Raid Noises in Psychotherapy', *British Medical Journal*, 2(4204): 158–9.

'Protests Diminish Bikini Goats' Rites' (22 July 1946), *New York Times*: 1.

Shephard, B. (2000), *A War of Nerves*, London: Random House.

Shurcliff, W. A. (1947), *Bombs at Bikini: The Official Report of Operation Crossroads*, New York: William H. Wise and Co.

Weisgall, J. M. (1994), *Operation Crossroads: The Atomic Tests at Bikini Atoll*, Annapolis, MD: Naval Institute Press.

HOW TO SURVIVE AN ATOMIC BOMB

'2 Boys and 35 Cows End Test in Shelter' (21 August 1963), *New York Times*: 15.

'Atom-Blasted Frozen Food Tastes Fine, Say Experts' (11 May 1955), *The Spokesman-Review*: 3.

Brinkley, B. (15 April 1949), 'Zoo Ready for Biggest Day of Year', *Washington Post*: B12.

Clarke, W.C. (January 1951), 'VD Control in Atom-Bombed Areas', *Journal of Social Hygiene* 37(1): 3–7.

Gerstell, R. (1950), *How to Survive an Atomic Bomb*, Washington, DC: Combat Forces Press.

Hanifan, D. T. (1963), *Physiological and Psychological Effects of Overloading Shelters*, Santa Monica, CA: Dunlap and Associates, Inc.

'Infant Find War, Open-Air Atom Survival Test Is Too Tough for Him' (7 November 1954), *New York Times*: 48.

Miles, M. (7 May 1955), 'Atom City Shows Few Could Have Survived', *Los Angeles Times*: 1.

'Pig 311', Record Unit 365, Series 4, Box 25, Folder 1, in Animal
    Information Files, 1855–1986 and undated, National Zoological
    Park, Office of Public Affairs.
Pinkowski, E. (1947), 'Imperishable Pig of Bikini is Still Subject of
    Debate', *Waterloo Sunday Courier*: 34.
Powers, G. (11 August 1951), 'Patty, the Atomic Pig', *Collier's* 128(6):
    24–5, 54.
Rose, K. D. (2001), *One Nation Underground: The Fallout Shelter in
    American Culture*, New York: New York University Press.
Vernon, J. A. (1959), *Project Hideaway: A Pilot Feasibility Study of Fallout
    Shelters for Families*, Princeton, NJ: Princeton University.
'What Science Learned at Bikini: Latest Report on the Results' (11 August
    1947), *Life* 23(6): 74–88.
'Woman Offers to Sit Thru Big A-bomb Test' (29 April 1955), *Chicago
    Daily Tribune*: 1.

NUKE THE MOON!

Arthur, C. (10 July 1999), 'Soviets Planned to Nuke the Moon', *Hamilton
    Spectator*: D4.
'Eerie Spectacle in Pacific Sky' (20 July 1962), *Life* 53(3): 26–33.
Dupont, D. G. (June 2004), 'Nuclear Explosions in Orbit', *Scientific
    American* 290(6): 100–7.
Ehricke, K. A., & G. Gamow (June 1957), 'A Rocket around the Moon',
    Scientific American, 196(6): 47–53.
Goddard, R. H. (26 August 1920), 'A Method of Reaching Extreme
    Altitudes', *Nature*, 2652(105): 809–11.
Krulwich, R. (1 July 2010), 'A Very Scary Light Show: Exploding H-
    Bombs in space', NPR. Available online at:
    http://www.npr.org/templates/story/story.php?storyId=128170775.
O'Neill, D. (1989), 'Project Chariot: How Alaska Escaped Nuclear
    Excavation', *Bulletin of the Atomic Scientists*, 45(10): 28–37.
Reiffel, L. (19 June 1959), *A Study of Lunar Research Flights*, Vol. 1,
    Kirtland Air Force Base, NM: Air Force Special Weapons Center.
Sullivan, W. (5 November 1957), 'Scientists Wonder if Shot Nears
    Moon', *New York Times*: 1.
Ulivi, P., with D. M. Harland (2004), *Lunar Exploration: Human Pioneers
    and Robotic Surveyors*, New York: Springer-Verlag.
Valente, J. (22 April 1991), 'Hate Winter? Here's a Scientist's Answer:
    Blow Up the Moon', *Wall Street Journal*: A1.

Vittitoe, C. N. (June 1989), *Did High-altitude EMP Cause the Hawaiian Streetlight Incident?*, Albuquerque, NM: Sandia National Laboratories.

THE INCREDIBLE ATOMIC SPACESHIP

Brownlee, R. R. (June 2002), 'Learning to Contain Underground Nuclear Explosions'. Available online at: http://nuclearweaponarchive.org/Usa/Tests/Brownlee.html.

Dyson, F. J. (19 July 1965), 'Death of a Project', *Science* 149(3680): 141–4.

Dyson, G. (2002), *Project Orion: The True Story of the Atomic Spaceship*, New York: Henry Holt and Co.

Three: Deceptive Ways

MEN FIGHT FOR SCIENCE'S SAKE

Buckhout, R. (1974), 'Eyewitness Testimony', *Scientific American*, 231(6): 23–31.

Buckhout, R., D. Figueroa, & E. Hoff (1975), 'Eyewitness Identification: Effects of Suggestion and Bias in Identification from Photographs', *Bulletin of the Psychonomic Society*, 6(1): 71–4.

Cattell, J. M. (1895), 'Measurements of the Accuracy of Recollection', *Science*, 2(49): 761–6.

Greer, D. S. (1971), 'Anything but the Truth? The Reliability of Testimony in Criminal Trials', *British Journal of Criminology*, 11(2): 131–54.

Jaffa, S. (1903), 'Ein psychologisches Experiment im kriminalistischen Seminar der Universität Berlin', *Beiträge zur Psychologie der Aussage*, 1(1): 79–99.

Marston, W. M. (1924), 'Studies in Testimony', *Journal of the American Institute of Criminal Law and Criminology*, 15(1): 5–31.

Münsterberg, H. (1917), *On the Witness Stand: Essays on Psychology and Crime*, Garden City, NY: Doubleday, Page & Company.

Von Liszt, F. (1902), 'Strafrecht und Psychologie', *Deutsche Juristen-Zeitung*, 7: 16–18.

THE PSYCHOLOGIST WHO HID BENEATH BEDS

Henle, M., & M. B. Hubbell (1938), '"Egocentricity" in Adult Conversation', *Journal of Social Psychology*, 9(2): 227–34.

Landis, M. H., & H. E. Burtt (1924), 'A Study of Conversations', *Journal of Comparative Psychology*, 4: 81–9.

'Mind-bending Disclosures' (15 August 1977), *Time*: 9.

Moore, H. T. (1922), 'Further Data Concerning Sex Differences', *Journal of Abnormal Psychology and Social Psychology*, 17(2): 210–14.

Stein, M. L. (1974), *Lovers, Friends, Slaves: The Nine Male Sexual Types*, New York: Berkley Publishing Corp.

THE METALLIC METALS ACT

Collett, P., & G. O'Shea (1976), 'Pointing the Way to a Fictional Place: A Study of Direction Giving in Iran and England', *European Journal of Social Psychology* 6(4): 447–58.

Gill, S. (14 March 1947), 'How Do You Stand on Sin?' *Tide*: 72.

Hartley, E. (1946), *Problems in Prejudice*, New York: Octagon Books.

Hawkins, D. I, & K. A. Coney (1981), 'Uninformed Response Error in Survey Research', *Journal of Marketing Research*, 18(3): 370–4.

Payne, S. L. (1950), 'Thoughts about Meaningless Questions', *Public Opinion Quarterly*, 14(4): 687–96.

Schuman, H., & S. Presser (1979), 'The Assessment of "No Opinion" in Attitude Surveys', *Sociological Methodology* 10: 241–75.

A ROOMFUL OF STOOGES

Asch, S. E. (November 1955), 'Opinions and Social Pressure', *Scientific American*, 193(5): 31–5.

Hall, R. L. (1958), 'Flavor Study Approaches at McCormick & Company, Inc.', in *Flavor Research and Food Acceptance*, New York: Reinhold Publishing Corp., 224–40.

Korn, J. H. (1997), *Illusions of Reality: A History of Deception in Social Psychology*, Albany: State University of New York Press.

Latané, B., & J. M. Darley (1970), *The Unresponsive Bystander: Why Doesn't He Help?*, Englewood Cliffs, NJ: Prentice-Hall, Inc.

Milgram, S. (1974), *Obedience to Authority: An Experimental View*, New York: Harper & Row.

THUD!

Rosenhan, D. L. (1973), 'On Being Sane in Insane Places', *Science*, 179(4070): 250–8.

——— (1975), 'The Contextual Nature of Psychiatric Diagnosis', *Journal of Abnormal Psychology*, 84(5): 462–74.

Spitzer, R. L. (1975), 'On Pseudoscience in Science, Logic in Remission, and Psychiatric Diagnosis: A Critique of Rosenhan's "On Being Sane in Insane Places"', *Journal of Abnormal Psychology*, 84(5): 442–52.

THE SEDUCTIVE DR FOX

Kaplan, R. M. (1974), 'Reflections on the Doctor Fox Paradigm', *Journal of Medical Education*, 49(3): 310–12.

Naftulin, D. H., J. E. Ware Jr., & F. A. Donnelly (1973), 'The Doctor Fox Lecture: A Paradigm of Educational Seduction', *Journal of Medical Education*, 48(7): 630–5.

Ware, J. E. Jr., & R. G. Williams (1975), 'The Dr Fox Effect: A Study of Lecturer Effectiveness and Ratings of Instruction', *Journal of Medical Education*, 50(2): 149–56.

Four: Monkeying Around

THE MAN WHO TALKED TO MONKEYS

'Garner Found Ape that Talked to Him' (6 June 1919), *New York Times*: 13.

Garner, R. L. (1892), 'A Monkey's Academy in Africa', *New Review*, 7: 282–92.

——— (1892), *The Speech of Monkeys*, London: William Heinemann.

——— (1896), *Gorillas & Chimpanzees*, London: Osgood, McIlvaine & Co.

——— (1900), *Apes and Monkeys: Their Life and Language*, Boston: Ginn & Company.

Radick, G. (2007), *The Simian Tongue: The Long Debate about Animal Language*, Chicago: University of Chicago Press.

Schmeck, H. M. (28 November 1980), 'Studies in Africa Find Monkeys Using Rudimentary Language', *New York Times*: 1.

Tickell, S. R. (1864), 'Notes on the Gibbon of Tenasserim, *Hylobates lar*', *Journal of the Asiatic Society of Bengal*, 33: 196–9.

CHIMPANZEE BUTLERS AND MONKEY MAIDS

'A Monkey College to Make Chimpanzees Human' (23 August 1924), *Hamilton Evening Journal*: 18.

'A Monkey with a Mind' (January 1910), *New York Times*: SM7.

'Can Science Develop Monkeys into Useful Men?' (7 May 1916),
     Washington Post: MT5.
Etkind, A. (2008), 'Beyond Eugenics: The Forgotten Scandal of
     Hybridizing Humans and Apes', *Studies in History and Philosophy of
     Biological and Biomedical Sciences*, 39: 205–10.
'French Doctors Experimenting with a Village of Tame Apes' (9
     November 1924), *Pittsburgh Press*: 111–12, 116.
Furness, W. H. (1916), 'Observations on the Mentality of Chimpanzees
     and Orang-utans', *Proceedings of the American Philosophical Society*,
     55(3): 281–90.
Goldschmidt, J. (30 March 1916), 'An Ape Colony to Suffer for
     Mankind', *New York Times*: 12.
'How Clothes Influence Monkeys' (2 July 1911), *Sandusky Register*: 1.
'If Science Should Develop Apes into Useful Workers' (14 May 1916),
     Washington Post: MT5.
Köhler, W. (1925), *The Mentality of Apes*, New York: Harcourt, Brace &
     Company.
'Monkeys Pick Cotton: Three Hundred of Them in Use on a Plantation'
     (5 February 1899), *Los Angeles Times*: 1.
Rossiianov, K. (2002), 'Beyond Species: Il'ya Ivanov and His Experiments
     on Cross-Breeding Humans with Anthropoid Apes', *Science in
     Context*, 15(2): 277–316.
Teuber, M. L. (1994), 'The Founding of the Primate Station, Tenerife,
     Canary Islands', *American Journal of Psychology*, 107(4): 551–81.
Witmer, L. (15 December 1909), 'A Monkey with a Mind', *Psychological
     Clinic*, 3(7): 179–205.

THE MAN WHO MADE APES FIGHT

Engels, J. (3 September 2004), 'Boos op alles en iedereen', *Trouw*: 16.
Hebb, D. O. (1946), 'On the Nature of Fear', *Psychological Review*, 53:
     259–76.
Hendriks, E. (28 August 2004), 'Zestig jaar tegen Niko en de anderen',
     De Volkskrant: W3.
Kortlandt, A. (1962), 'Chimpanzees in the Wild', *Scientific American*,
     206(5): 128–38.
——— (1966), 'Experimentation with Chimpanzees in the Wild', in
     Starck, D., R. Schneider, & H. J. Kuhn, eds., *Neue Ergebnisse der
     Primatologie*, Stuttgart: Gustav Fischer Verlag, 208–24.

———— (1980), 'How Might Early Hominids Have Defended Themselves Against Large Predators and Food Competitors?', *Journal of Human Evolution*, 9: 79–112.

Röell, D. R. (2000), *The World of Instinct: Niko Tinbergen and the Rise of Ethology in the Netherlands (1920–1950)*, Assen, Netherlands: Van Gorcum.

van Hooff, J. (2000), 'Primate Ethology and Socioecology in the Netherlands', in Strum, S. C., & L. M. Fedigan, eds., *Primate Encounters: Models of Science, Gender, and Society*, Chicago: Chicago University Press, 116–37.

THE MYSTERY OF THE MISSING MONKEY ORGASM

Burton, F. D. (1971), 'Sexual Climax in Female *Macaca mulatta*', *Proceedings of the 3rd International Congress of Primatology*, 3: 180–91.

Chevalier-Skolnikoff, S. (1974), 'Male-Female, Female-Female, and Male-Male Sexual Behavior in the Stumptail Monkey, with Special Attention to the Female Orgasm', *Archives of Sexual Behavior*, 3(2): 95–116.

Dewsbury, D.A. (2006), *Monkey Farm: A History of the Yerkes Laboratories of Primate Biology, Orange Park, Florida, 1930–1965*. Lewisburg: Bucknell University Press.

Ford, C. S., & F. A. Beach (1951), *Patterns of Sexual Behavior*, New York: Harper.

Hamilton, G. V. (1914), 'A Study of Sexual Tendencies in Monkeys and Baboons', *Journal of Animal Behavior*, 4(5): 295–318.

Haraway, D. (1989), *Primate Visions: Gender, Race, and Nature in the World of Modern Science*, New York: Routledge.

Hrdy, S. B. (1999), *The Woman that Never Evolved*, Cambridge, MA: Harvard University Press.

Yerkes, R. M. (1939), 'Sexual Behavior in the Chimpanzee', *Human Biology*, 11(1): 78–111.

Zumpe, D., & R. P. Michael (1968), 'The Clutching Reaction and Orgasm in the Female Rhesus Monkey (*Macaca mulatta*)', *Journal of Endocrinology*, 40(1): 117–23.

THE CHIMP WHO GREW UP HUMAN

Brown, E. (3 February 1974), 'She's Mother to a Kitten for Science's Sake', *Los Angeles Times*: B1.

Carter, J. (April 1981), 'A Journey to Freedom', *Smithsonian*, 12(1): 90–101.

———— (June 1988), 'Freed from Keepers and Cages, Chimps Come of Age on Baboon Island', *Smithsonian*, 19(3): 36–49.

Douthwaite, J. V. (2002), *The Wild Girl, Natural Man, and the Monster: Dangerous Experiments in the Age of Enlightenment*, Chicago: University of Chicago Press.

Gorner, P. (8 January 1976), 'The Birds, Bees, and Lucy', *Chicago Tribune*: B1.

Hess, E. (2008), *Nim Chimpsky: The Chimp Who Would Be Human*, New York: Bantam Dell.

Kellogg, W. N., & L. A. Kellogg (1933), *The Ape and the Child: A Study of Environmental Influence upon Early Behavior*, New York: McGraw-Hill Book Company, Inc.

Peterson, D. (1995), *Chimpanzee Travels: On and Off the Road in Africa*, Reading, MA: Addison-Wesley.

Temerlin, M. K. (1975), *Lucy: Growing Up Human. A Chimpanzee Daughter in a Psychotherapist's Family*, Palo Alto, CA: Science and Behavior Books, Inc.

Five: Do-It-Yourselfers

HARD TO SWALLOW

Carlson, A. J. (1918), 'Contributions to the Physiology of the Stomach. XLV. Hunger, Appetite and Gastric Juice Secretion in Man During Prolonged Fasting (Fifteen Days)', *American Journal of Physiology*, 45: 120–46.

Cathrall, I. (1802), 'Memoir on the Analysis of Black Vomit', *Transactions of the American Philosophical Society*, 5: 117–38.

Elmore, J. G. (September 1994), 'Joseph Goldberger: An Unsung Hero of American Clinical Epidemiology', *Annals of Internal Medicine*, 121(5): 372–75.

Goldberger, J. (1 April 1917), 'The Transmissibility of Pellagra: Experimental Attempts at Transmission to the Human Subject', *Southern Medical Journal*, 10(4): 277–86.

Grassi, B. (1888), 'Weiteres zur frage der Ascarisentwickelung', *Centralblatt für Bakteriologie und Parasitenkunde*, 3(24): 748–9.

'Has an Appetite for Hardware, Glass and Gravel' (30 April 1933),
   Milwaukee Sentinel: 47.

Hoelzel, F. (1930), 'The Rate of Passage of Inert Materials through the
   Digestive Tract', *American Journal of Physiology*, 92: 466–97.

Koino, S. (15 November 1922), 'Experimental Infections on Human
   Body with Ascarides', *Japan Medical World*, 2(2): 317–20.

'M. Guyon's Experiments on the Contagion of Yellow Fever' (1823),
   Quarterly Journal of Foreign and British Medicine and Surgery, 5: 443–4.

Prescott, F. (February 1930), 'Spallanzani on Spontaneous Generation
   and Digestion', *Proceedings of the Royal Society of Medicine*, 23(4):
   495–510.

La Roche, R. (1855), *Yellow Fever Considered in its Historical, Pathological,
   Etiological, and Therapeutical Relations*, Philadelphia: Blanchard and
   Lea.

'On the Conversion of Cysticercus Cellulosae into *Taenia solium* by Dr.
   Kuchenmeister' (1861), *American Journal of the Medical Sciences*,
   41: 248–9.

Spallanzani, L. (1784), *Dissertations Relative to the Natural History of
   Animals and Vegetables*, Vol. 1, London: J. Murray.

White, A. S., R. D. Godard, C. Belling, V. Kasza, & R. L. Beach (2010),
   'Beverages Obtained from Soda Fountain Machines in the U.S.
   Contain Microorganisms Including Coliform Bacteria', *International
   Journal of Food Microbiology*, 137(1): 61–6.

THIS WILL BE EXTREMELY PAINFUL

Baerg, W. J. (1923), 'The Effects of the Bite of *Latrodectus mactans* Fabr',
   Journal of Parasitology, 9(3): 161–9.

Barnes, J. H. (13 June 1964), 'Cause and Effect in Irukandji Stingings',
   Medical Journal of Australia, 1(24): 897–904.

Blair, A. W. (1934), 'Spider Poisoning: Experimental Study of the Effects
   of the Bite of the Female *Latrodectus mactans* in Man', *Archives of
   Internal Medicine*, 54(6): 831–43.

Conniff, R. (June 2003), 'Stung: How Tiny Little Insects Get Us to Do
   Exactly as They Wish', *Discover*, 24(6): 67–70.

Dye, S. F., G. L. Vaupel, & C. C. Dye (1998), 'Conscious Neurosensory
   Mapping of the Internal Structures of the Human Knee without
   Intraarticular Anesthesia', *American Journal of Sports Medicine*,
   26(6): 773–7.

Head, H., & W. H. R. Rivers (November 1908), 'A Human Experiment in Nerve Division', *Brain*, 31: 323–450.

Kellgren, J. H. (1939), 'On the Distribution of Pain Arising from Deep Somatic Structures with Charts of Segmental Pain Areas', *Clinical Science*, 4(1): 35–46.

Peterson, D. F., & A. M. Brown (1973), 'Functional Afferent Innervation of Testis', *Journal of Neurophysiology*, 36(3): 425–33.

Ryan, C. (1995), *The Pre-Astronauts: Manned Ballooning on the Threshold of Space*, Annapolis: Naval Institute Press.

Schmidt, J. O. (1990), 'Hymenopteran Venoms: Striving Toward the Ultimate Defense against Vertebrates', in Evans, D.L., & J.O. Schmidt, eds., *Insect Defenses: Adaptive Mechanisms and Strategies of Prey and Predators*, Albany: State University of New York Press, 387–419.

Woollard, H. H., & E. A. Carmichael (1933), 'The Testis and Referred Pain', *Brain*, 56(3): 293–303.

ADVENTURES IN SELF-SURGERY

Altman, L. K. (1987), *Who Goes First? The Story of Self-Experimentation in Medicine*, New York: Random House.

Bier, A. (1899), 'Versuche über Cocainisierung des Rückenmarks', *Deutsche Zeitschrift für Chirurgie*, 51: 361–9.

Bishop, W. J. (1961), 'Some Historical Cases of Auto-Surgery', *Proceedings of the Scottish Society of the History of Medicine*, Session 1960–1: 23–32.

'Doctor Operates on Himself: Astonishing Experiment' (1 April 1912), *Wellington Evening Post*: 7.

'Dr. Evan Kane Dies of Pneumonia at 71' (2 April 1932), *New York Times*: 23.

Freud, S. (1974), *Cocaine Papers*, edited by Anna Freud, New York: New American Library.

Frost, J. G., & C. G. Guy (16 May 1936), 'Self-Performed Operations: With Report of a Unique Case', *Journal of the American Medical Association*, 106(20): 1708–10.

'German Doctor Reports he Removed His Own Appendix' (10 December 1948), *Los Angeles Times*: 12.

Gille, M. (1933), 'L'auto-chirurgie', *L'Echo Medical Du Nord*, 37: 45–8.

Kane, E. O. (March 1921), 'Autoappendectomy: A Case History', *International Journal of Surgery*, 34(3): 100–2.

Knapp, H. (1885), *Cocaine and its Use in Ophthalmic and General Surgery*, New York: G. P. Putnam's Sons.

Murphy, L. J. T. (1969), 'Self-Performed Operations for Stone in the Bladder', *British Journal of Urology*, 41(5): 515–29.

Rogozov, V., & N. Bermel (2009), 'Autoappendicectomy in the Antarctic', *British Medical Journal*, 339: 1420–2.

Streatfeild, D. (2002), *Cocaine: An Unauthorized Biography*, New York: Thomas Dunne Books.

'Suicide of Dr. Horace Wells, of Hartford Connecticut, U.S.' (31 May 1848), *Providence Medical and Surgical Journal*, 12(11): 305–6.

'Surgeon Operates on Self: Brazilian Professor Astonishes Students by Unique Experiment' (6 January 1929), *New York Times*: 33.

KILLING YOURSELF FOR SCIENCE

Bacon, F. (1623), *Historia Vitae et Mortis*, London: M. Lownes.

'Doctor Scribbles Narrative of Own Death by Narcotic' (27 November 1936), *Los Angeles Times*: 1, 7.

Fleichmann, M. (1832), 'Des différents genres de mort par strangulation', *Annales d'hygiène publique et de médecine légale*, 8(1): 416–41.

'Freezing Affects Mind First; Initiative and Modesty Lost' (17 October 1936), *Science News-Letter*, 30(810): 252.

Hammond, G. M. (1882), 'On the Proper Method of Executing the Sentence of Death by Hanging', *Medical Record*, 22: 426–8.

Holstein, B. (2009), 'The Mysterious Disappearance of Ettore Majorana', *Journal of Physics: Conference Series*, 173: 012019.

Metchnikoff, O. (1921), *Life of Elie Metchnikoff: 1845–1916*, Boston: Houghton Mifflin Company.

Minovici, N. S. (1905), *Étude sur la pendaison*, Paris: A. Maloine.

Zaslavskii, O.B. (2006), 'Ettore Majorana: Quantum Mechanics of Destiny', *Priroda* 11: 55–63. http://arXiv.org/pdf/physics/0605001.

ACKNOWLEDGEMENTS

This book wouldn't have been possible without the assistance and support of many people. In particular, I'm grateful to Jon Butler and Natasha Martin at Macmillan for their guidance and patience during the editing process.

My writing group partners, Sally Richards and Jennifer Donohue, kept my feet to the fire by reading and critiquing drafts of the manuscript every week. I'm pretty sure the book would have taken me twice as long to finish if I hadn't known that I had to get pages to them every week on time.

This book was very research intensive, so I owe a great debt to all the librarians (especially those at the University of California San Diego library and the La Mesa Public Library) who unfailingly went out of their way to help me whenever I showed up searching for various strange and obscure titles.

And then there are all the usual suspects who provided the moral support that kept me going through the months of research and writing: my awesome parents (#1 Fan and #1 Handyman), the Palo Alto crowd (Kirsten, Ben, Pippa, and Astrid), Charlie 'master chef' Curzon, Ted 'coffee break' Lyons, Boo the high-maintenance cat, my father-in-law John Walton, who sent good thoughts from South Africa, and most of all my wife Beverley, who kept me oriented toward sanity as I immersed myself deep in the bizarre world of mad scientists.